STEREOCHEMISTRY AND ITS APPLICATION IN BIOCHEMISTRY

STEREOCHEMISTRY AND ITS APPLICATION IN BIOCHEMISTRY

The Relation Between Substrate Symmetry and Biological Stereospecificity

WILLIAM L. ALWORTH

Department of Chemistry
Tulane University
New Orleans, Louisiana

WILEY-INTERSCIENCE, a Division of John Wiley & Sons, Inc.

New York • London • Sydney • Toronto

Copyright © 1972, by John Wiley & Sons, Inc.

All rights reserved. Published simultaneously in Canada.

No part of this book may be reproduced by any means, nor transmitted, nor translated into a machine language without the written permission of the publisher.

Library of Congress Cataloging in Publication Data

Alworth, William L. 1939–
    Stereochemistry and its application in biochemistry.

    Bibliography: p.
    1. Biological chemistry.   2. Stereochemistry.
I. Title.

QD415.A43   574.1'92'01541223   76–39308
ISBN 0–471–02518–6

Printed in the United States of America.

10 9 8 7 6 5 4 3 2 1

To Lois, who still does not understand biological
stereospecificity—but who was generally understanding

PREFACE

The concept that molecules of biological origin may be optically active is introduced early in organic chemistry courses. The beginning biochemistry student therefore readily accepts the fact that biological systems manifest biological stereospecificity and differentiate between stereoisomeric substrate molecules. The idea that biological systems evidence additional forms of biological stereospecificity and differentiate between the chemically like, paired groups of individual substrate molecules is more difficult for the beginning biochemistry student to accept. This book describes substrate symmetry and stereochemistry in a systematic manner. The presentation then emphasizes and illustrates how these molecular properties govern biological stereospecificity between chemically like, paired groups. Since the principles underlying this phenomenon have often been misunderstood by established biochemists, it is hoped that this systematic introduction will aid the understanding of practicing biochemists, as well as leading to a new generation of more knowledgeable students in the field.

The presentation is designed so that the introductory biochemistry student, with the aid of a molecular model set, can gain an insight into the stereochemical principles that are involved in molecular interactions within biological systems. Experience indicates that the average biology student will have difficulty mastering the sections dealing with molecular symmetry and may have to engage in extensive model building. The average chemistry student, on the other hand, may find that much of the discussion of symmetry and stereochemistry is simply a review. The biology student, however, may be familiar with the biochemical reactions cited to illustrate aspects of enzyme stereospecificity, whereas the chemistry major will have to consult a biochemical text to properly place these reactions within general metabolic schemes.

Whatever their backgrounds, all students will profit by considering this study of molecular symmetry and biological stereospecificity as a vital part of an introduction to the field of biochemistry. Biological stereospecificity between chemically like, paired groups is involved in many of the fundamental pathways of intermediary metabolism, including the TCA cycle, sterol biosynthesis, and ketose–aldose interconversions. The "classical" examples of biological stereospecificity between chemically like, paired groups, which should be familiar to every student of biochemistry, are described and analyzed in this book. In addition, more recent, exciting experimental observations of biological stereospecificity between the chemically like methyl hydrogens of chiral acetate and pyruvate substrates are presented and explained.

After studying this book, the student will be able to fully comprehend recent experimental results, dealing with enzymatic stereospecificity, which are dependent on a working knowledge of molecular symmetry.

He will also be able to confidently predict which additional enzymatic reactions may be eventually demonstrated to differentiate between chemically like, paired groups.

I acknowledge the helpful comments and suggestions of O. E. Weigang, Jr., and J. Hamer, M. V. Keenan, and C. B. Frederick, who kindly consented to read an early draft of this book.

WILLIAM L. ALWORTH

November 1971
New Orleans, Louisiana

CONTENTS

THE GENTLEMAN OF SHALOTT

Which eye's his eye?
Which limb lies
next the mirror?
For neither is clearer
nor a different color
than the other,
nor meets a stranger
in this arrangement
of leg and leg and
arm and so on.
To his mind
it's the indication
of a mirrored reflection
somewhere along the line
of what we call the spine.

He felt in modesty
his person was
half looking-glass,
for why should he
be doubled?
The glass must stretch
down his middle,
or rather down the edge.
But he's in doubt
as to which side's in or out
of the mirror.
There's little margin for error
but there's no proof, either.
And if half his head's reflected,
thought, he thinks, might be affected.

But he's resigned
to such economical design.
If the glass slips
he's in a fix—
only one leg, etc. But
while it stays put
he can walk and run
and his hands can clasp one
another. The uncertainty
he says he
finds exhilarating. He loves
that sense of constant re-adjustment.
He wishes to be quoted as saying at present:
"Half is enough."

Elizabeth Bishop—The Complete Poems
Farrar, Straus and Giroux (1969)

STEREOCHEMISTRY
AND ITS APPLICATION
IN BIOCHEMISTRY

Chapter I

HISTORICAL DEVELOPMENT OF THE CONCEPT OF BIOLOGICAL STEREOSPECIFICITY

Soon after the development of experimental techniques for producing beams of plane-polarized light, the French physicist Biot observed that when such beams were passed through solutions containing chemicals of biological origin, the plane of the polarization was rotated. In 1848 Pasteur succeeded in linking this "optical activity" of a biological compound in solution with the phenomenon of hemihedral crystals of the compound in the solid phase. In his classical studies on the tartaric acids, Pasteur crystallized the sodium ammonium salt of racemic tartaric acid under conditions (below 27°, from dilute solutions) where discernible hemihedral tartaric acid salt crystals were formed, some with the faces arranged to the right and some arranged to the left. Crystals that display unequal development or modification of related parts and thus lack points or planes of symmetry are said to be *hemihedral*. As Figure I-1 reveals, such hemihedral crystals may exist in *enantiomeric* forms; that is, they may be related as object and nonsuperimposable mirror image. By carefully separating the enantiomeric crystals with the aid of a microscope, Pasteur was able to demonstrate that the crystals hemihedral in the right-hand sense, when dissolved in solution, rotated the plane of polarized light to the right, whereas solutions of the crystals hemihedral in the left-hand sense rotated the plane of polarized light to the left. When Pasteur isolated and purified the free tartaric acids from the separate salt solutions, he found one acid to be identical to the previously characterized dextrorotatory tartaric acid [*d*-tartaric acid or (+)-tartaric acid: rotation to the right] and the other acid to be the previously unknown levorotatory tartaric acid [*l*-tartaric acid or (−)-tartaric acid: rotation to the same extent, but to the left]. Thus Pasteur had successfully accomplished the first recorded

1

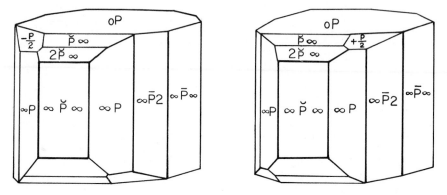

Figure I-1. Hemihedral crystals of sodium ammonium tartrate. This figure is based upon Pasteur's original descriptions and was first published in Kekulé's *Lehrbuch der Organischen Chemie*, 1866; cf., Ref. 1.

resolution of a *racemic** mixture (a *dl*,1:1 mixture) into its dextrorotatory and levorotatory optically active components.

The dextrorotatory and levorotatory tartaric acids have the same melting point, solubilities, dissociation constant, density, and general chemical behavior. They are identical except in that they rotate the plane of polarized light in opposite directions (but to the same extent), and in that they form crystals that are hemihedral in the opposite sense (under special conditions). Pasteur related these two phenomena and concluded that tartaric acid must possess an asymmetry within its fundamental molecular geometry. In 1860 Pasteur concluded:

(1) The molecule of tartaric acid, whatever else it may be, is asymmetric, and in such a way that the image [mirror] is not superposable. (2) The molecule of the left tartaric acid is formed by the exact inverse group of atoms. And by what characters shall we recognize the existence of molecular asymmetry? On the one hand by non-superposable hemihedry, on the other, and more especially, by the rotatory optical property when the substance is in solution.[2]

* Racemic acid was studied, characterized, and named by Gay-Lussac in 1826. He recognized that, although it possessed differentiable chemical properties, it had the same empirical formula ($C_4H_6O_6$) as tartaric acid. In 1830 Berzelius proposed the term *isomers* to describe such chemical compounds. Until the work of Pasteur it was not appreciated that the optically inactive racemic acid was actually a 1:1 mixture of *d* and *l*-tartaric acids. Racemic compounds as exemplified by *dl*-tartaric acid, although optically inactive, are potentially resolvable into their active *d* and *l* component parts. They are to be contrasted with other possible stereoisomeric forms, exemplified by *meso*-tartaric acid, which are inherently optically inactive (cf. Chapter III).

Pasteur recognized that the phenomena of hemihedral enantiomeric crystals and of optical rotation were related. He also clearly differentiated between the optical activity of hemihedral quartz crystals and that of solutions of natural organic compounds.

Permit me to illustrate roughly, although with essential accuracy, the structure of quartz and of the natural organic products. Imagine a spiral stair whose steps are cubes, or any other objects with superposable images. Destroy the stair and the asymmetry will have vanished. The asymmetry of the stair was simply the result of the mode of arrangement of the component steps. Such is quartz. The crystal of quartz is the stair complete. It is hemihedral. It acts on polarized light in virtue of this. But let the crystal be dissolved, fused, or have its physical structure destroyed in any way whatever; its asymmetry is suppressed and with it all action on polarized light, as it would be, for example, with a solution of alum, a liquid formed of molecules of cubic structure distributed without order.

Imagine, on the other hand, the same spiral stair to be constructed with irregular tetrahedra for steps. Destroy the stair and the asymmetry will still exist, since it is a question of a collection of tetrahedra. They may occupy any positions whatsoever, yet each of them will none the less have an asymmetry of its own. Such are the organic substances in which all the molecules have an asymmetry of their own, betraying itself in the form of the crystal. When the crystal is destroyed by solution, there results a liquid active towards polarized light, because it is formed of molecules, without arrangement, it is true, but each having an asymmetry in the same sense, if not of the same intensity in all directions.[3]

Although Pasteur, in 1860, was emphasizing that the optical activity of organic solutions was determined by a molecular asymmetry which produced nonsuperimposable mirror image structures—and even describing an optically active organic solution as a collection of irregular tetrahedra—his insight was far ahead of organic structural theory. Not until 1874 was a theory of organic molecular structure proposed which was consistent with the comprehension demonstrated in Pasteur's statements. In that year, two young chemists independently and nearly simultaneously suggested that the four valences of the carbon atom were not planar but were directed into three-dimensional space.[4] A compound containing a carbon substituted with four different groups (C_{abcd}) would therefore be capable of existing in two distinctly different nonsuperimposable forms just as foreseen by Pasteur. The Frenchman Le Bel realized that such dissymmetry could exist if the four carbon valencies were directed to four noncoplanar points;[5] the Dutchman van't Hoff proposed a specific tetrahedral arrangement.[6] Although van't Hoff had no theoretical basis for proposing a tetrahedral bonding arrangement for carbon, his original structural formulas could still be used in a modern organic text. Today the organic chemist views a tetravalent carbon atom as situated in the center of a regular tetrahedron with four equivalent bonding orbitals composed of sp^3-hybridized atomic orbitals directed toward the vertices of the tetrahedron. As we discuss in Chapters II and III, this geometrical arrangement of a single C_{abcd} grouping is a sufficient (but not necessary) condition to produce enantiomeric molecules related as object and nonsuperimposable mirror image.

DEFINITION OF BIOLOGICAL STEREOSPECIFICITY

Since the tartaric acids are formed as by-products in the microbiological processes that produce wines, optical activity has always been associated with compounds of biological origin. The biochemistry student is usually aware that optical activity is one of the most characteristic results of a biochemically catalyzed process. He readily accepts that enzymes will differentiate between the enantiomeric forms of substrate molecules. In contrast, the introductory biochemistry student has difficulty in accepting that a biological stereospecific process will also differentiate between the *chemically like*, but *geometrically nonequivalent, paired groups of a single substrate molecule.* Actually, the latter aspect of biological stereospecificity is an easier concept to understand than is optical activity. It is initially more difficult to accept only because of the lack of systematic treatments of this feature of biological stereospecificity in introductory texts. The following chapters illustrate the very real relationship that exists between these *two aspects of biological stereospecificity.* By stressing one particular aspect of biological stereospecificity, the phenomenon of enzymatic discrimination between chemically like, paired groups, we hope that this book will aid in overcoming a deficiency of many introductory biochemical texts.

Before proceeding further, we should define more precisely the type of enzyme specificity we are describing here as "stereospecificity." Enzymes are readily divided into classes depending on the specific kind of reaction they catalyze. Thus there are oxidoreductases—catalyzing oxidation-reduction reactions, isomerases—catalyzing isomerizations, hydrolases—catalyzing hydrolytic reactions, and so on. The enzymatic property of catalyzing a specific type of chemical reaction is termed *reaction specificity.* This aspect of enzyme specificity is analogous to other catalytic processes and can be understood in terms of basic chemical concepts. Within the basic groupings of enzymes there are a number of subclasses, based on further reaction specificities. Within the group of enzymes catalyzing hydrolytic reactions there will be subclasses of enzymes that hydrolyze peptides—peptidases, phosphoric acid esters—phosphatases, carboxylic acid esters—esterases, and so on. Again, this aspect of enzyme specificity is readily understood in terms of the basic chemical differences in the reaction being catalyzed. Peptide (amide) bonds are chemically distinct from ester bonds, and the enzymatic reaction specificity is analogous to different chemical catalysis conditions that will preferentially hydrolyze certain classes of bonds. Even within the subclass of enzymes specifically hydrolyzing peptide (amide) bonds, however, there will be an additional preference for the peptide linkages of specific amide substrates. Thus pepsin has a marked specificity for peptide bonds adjacent to aromatic or dicarboxylic amino acid residues, and trypsin has a specificity for the peptide bonds involving the carboxyl groups of the basic amino acids arginine or lysine. When it is realized that enzymatic catalysis takes place within a three-dimensional active "pocket" or region on an enzyme surface, then this additional *substrate specificity* of enzymes can also be understood in terms of chemical differences. The productive interaction of a substrate with the enzymatic catalytic site will require favorable interactions of the enzyme and the substrate at several points in addition to that specific portion of the substrate undergoing reaction. * The chemical nature of the groups surrounding the reactive site of the substrate will determine whether favorable ionic, hydrophobic, and steric interactions with the total active pocket of the protein are possible.† In this way, for example, the chemical nature of the amino acid groups adjacent to the peptide bond will influence the effectiveness of the catalytic interactions and will lead to the observed substrate specificities of pepsin and trypsin. Similar factors lead to the observed specificity of glucokinase (ATP:glucose-6-phosphotransferase)‡, both for glucose as a specific hexose substrate and for the unique primary alcohol group of glucose as a specific phosphorylation site. Thus these types of enzymatic reaction and substrate specificity may be understood in terms of *definable chemical differences.*

In contrast, *enzymatic stereospecificity* refers to the ability of these remarkable catalysts to differentiate *between chemically like groups* on the basis of their *absolute geometrical differences.*§ Enzymatic stereospecificity is manifest in bio-

---

* Cf. J. D. Watson, *Molecular Biology of the Gene*, Chapter 4, W. A. Benjamin, 1965.

† Cf. Chapter V for a discussion of α-chymotrypsin substrate specificity.

‡ Enzymes are generally referred to here by trivial names, but when first mentioned, the systematic nomenclature established by the International Union of Biochemistry is also indicated.

§ Chemists have described the differentiated reactivities resulting from geometrical nonequivalence by two distinct terms—stereoselective and stereospecific. A chemical *stereoselective* reaction is defined as a reaction in which a predominance of one particular stereoisomer is either converted or formed. A chemical *stereospecific* reaction is one in which different stereoisomeric reactants yield contrasting stereoisomeric products.

In many of the biochemical examples cited in this book, these two aspects of differentiated stereoreactivity become thoroughly intertwined. For such reasons the distinctions between stereoselective and stereospecific processes are seldom useful in biochemical descriptions. Here a broad definition of biological stereospecificity is employed and then this term is applied to all aspects and effects of differentiated enzymatic stereoreactivity (see also pp. 178–179).

chemical systems in two ways. One aspect of biological stereospecificity is the differentiation between enantiomeric forms of substrate or product molecules. This aspect of biological stereospecificity is the more familiar and is directly related to optical activity in materials of biological origin. Recognition of this type of biological stereospecificity may therefore be traced to Biot's investigations with polarized light, made early in the nineteenth century. The second aspect of enzymatic stereospecificity is the differentiation between chemically like, paired groupings of an individual substrate or product molecule. As we shall see, recognition of this less familiar aspect of biological stereospecificity did not occur until the 1940s. These two stereochemical aspects of enzyme-substrate specificity can be illustrated by considering the examples in Figure I-2.

Considering the glycerol molecule as a potential enzymatic substrate, chemical differences allow us to predict that an enzyme will be able to distinguish the secondary alcohol group (on C-2) from either of the two primary alcohol functions (at C-1 and C-3). For example, oxidation of the secondary alcohol group leads to a ketone, whereas oxidation of either of the primary alcohol groups would lead to aldehyde products. Similarly, an enzymatic esterification of glycerol could be expected to differentiate the primary alcohol functions from the unique secondary alcohol position on the basis of chemical differences. Chemical esterifications often can lead to selective reaction at the less hindered primary alcohol position in preference to a

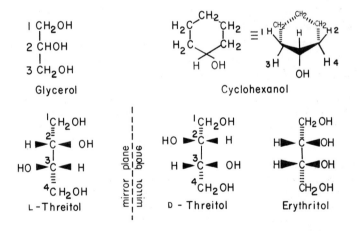

Figure I-2.   Potential substrates to illustrate enzymatic stereospecificity. *

* A standard stereochemical convention is used in the figures in this book. The solid lines represent chemical bonds lying in the plane of the page. Dashed lines are used to represent chemical bonds lying below or behind the plane of the page. The thickened, "wedged" lines are used to represent bonds that project above the plane of the page.

secondary or tertiary alcohol position. However, in terms of chemical functionality, even in terms of subtle reactivity differences imposed by the *relative positions* of other functional groups, the two primary alcohol functions at C-1 and C-3 of a glycerol molecule *are chemically alike*. Nevertheless, enzymes will catalyze the reaction of *one specific primary alcohol group of a glycerol substrate molecule owing to the absolute geometrical nonequivalence of these chemically like, paired alcohol functions*. Thus glycerol kinase (ATP:glycerol phosphotransferase) always phosphorylates one geometrically unique primary alcohol group of glycerol to produce exclusively L-α-glycerol phosphate —not the mixture of D and L-α-glycerol phosphates that would result from indiscriminate esterification at either of the two paired primary alcohol groups.

Consider the related case of cyclohexanol. If cyclohexanol were to serve as a substrate for a hypothetical dehydration enzyme that formed cyclohexene, which proton would be eliminated? Dehydration of cyclohexanol to cyclohexene involves the loss of the hydroxyl group and an adjacent proton. Chemical considerations alone would lead to the prediction that either one of the two chemically equivalent hydrogens *trans* to the hydroxyl (numbered 1 and 2 in Figure I-2) or one of the two *cis* hydrogens (numbered 3 and 4 in Figure I-2) would be removed. Various known chemically catalyzed reactions show a preference for either *syn* (elimination of *cis*-oriented groups) or *anti* (elimination of *trans*-oriented groups) elimination processes. Considering enzymatic stereospecificity, however, we would inquire additionally whether an enzyme catalyzing an *anti* elimination process could select between the chemically like, paired *trans* hydrogens 1 and 2. The answer, based on the rules to be developed in this book, is yes. If our hypothetical enzyme possessed the usual enzymatic stereospecificity, it would catalyze the elimination of only one of these two chemically like, paired *trans* hydrogens.

Finally, let us consider the case of D and L isomers of threitol. The arguments developed in the case of the glycerol molecule lead to the observation that chemical differences would permit differentiation of the alcohol groups at C-1 and C-4 (primary alcohol functions) from those alcohol groups at C-2 and C-3 (secondary alcohol functions). However, the related primary alcohol groups of D and of L-threitol are chemically alike, even with regard to the *relative* geometry of the adjacent chemical groups. D and L-threitol differ only with regard to the *absolute* three-dimensional orientation of related chemical groups. These two *stereoisomers* are related as object and nonsuperimposable mirror image. They are, therefore, *enantiomers*, just as the hemihedral tartaric acid crystals of Figure I-1. Enzymatic stereospecificity will again lead to a selective reaction with the primary alcohol functions of only *one* of the chemically like, enantiomeric threitol substrates because of the absolute geometrical nonequivalence. In an enzymatic process that displays a reaction specificity for a primary alcohol function due to chemical differences and a substrate stereospecificity for D-threitol due to absolute geometrical differences, we additionally want to know if enzymatic stereospecificity will permit differentiation between the *paired primary alcohol functions* at C-1 and C-4 of a reactive D-threitol substrate molecule. The answer is no. Based on the analyses to be developed in Chapters II and V, an enzyme could not differentiate between the paired primary alcohol groups of a D-threitol molecule. On the other hand, the principles of biological stereospecificity developed in Chapters II and V allow us to predict that an enzyme could differentiate between the paired primary alcohol groups of an erythritol molecule—if erythritol were the active substrate.

THE OGSTON EFFECT IN TCA METABOLISM

The long history of the recognition that biological stereospecificity results in differentiation between optical isomers is, of course, related to the development of techniques for observing and measuring optical rotation, early in the nineteenth century. Realization that biological stereospecificity *also* results in differentiation between chemically like, paired groups had to await the development of isotopic labeling methods. As techniques for producing significant amounts of the uncommon isotopic forms of the elements were not developed until the late 1930s, the history of the recognition of this latter aspect of biological stereospecificity is relatively short.

Figure I-3 outlines a portion of the familiar tricarboxylic acid (TCA) cycle. In 1941 Wood et al.,[7] using $^{13}CO_2$ and Evans and Slotin,[8] using $^{11}CO_2$, reported that the α-ketoglutarate formed by operation of the TCA cycle in the presence of the labeled CO_2 and pyruvate was labeled *solely* in the carboxyl group *adjacent* to the carbonyl (ketone) carbon. Both groups of workers were actually studying the fixation of carbon dioxide by pyruvate to form oxaloacetate in pigeon liver.

$$*CO_2 + CH_3C\overset{\|}{\underset{O}{}}-CO_2H \rightarrow HO_2C*-CH_2-C\overset{\|}{\underset{O}{}}-CO_2H$$

It was observed, however, that the resulting oxaloacetate was partially converted into α-ketoglutarate. Since we now realize that this process involves TCA cycle metabolism, the labeling pattern indicated in Figure I-3 must be explained.

At about the time the foregoing experimental results were obtained, H. A. Krebs was formulating the reactions of the TCA cycle.[9] Krebs had predicted

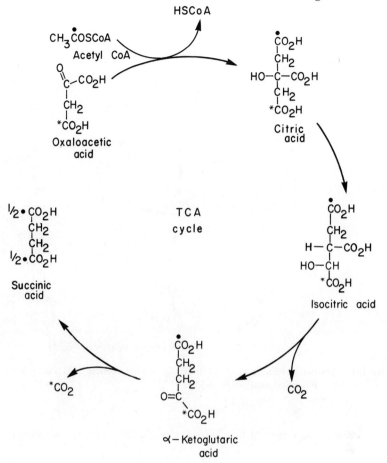

Figure I-3. Representation of the tricarboxylic acid (TCA) cycle. Carbon-labeling patterns within the TCA intermediates are indicated by asterisks and dots.

that isotopically labeled carbon dioxide, fixed into oxaloacetate, would result in the formation of isotopically labeled α-ketoglutarate as a result of the operation of the reactions outlined in Figure I-3. Krebs indicated, however, that this labeled oxaloacetate would lead to α-ketoglutarate molecules with isotopic labeling *distributed equally between both carboxyl groups*. Krebs predicted that this would occur because of the formation of two different labeled iso-citrate molecules as outlined in Figure I-4. The experimental results were clearly contrary to this part of the prediction. Evans and Slotin, and Wood et al. concluded that the formation of specifically labeled α-ketoglutarate in their pigeon liver systems could not have proceeded via citrate as proposed

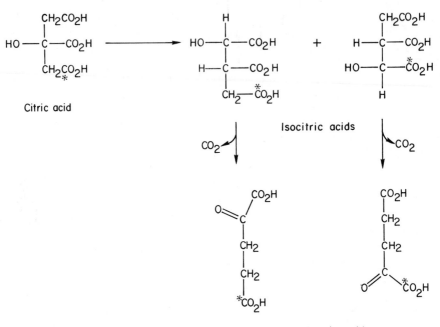

Figure I-4. Postulated formation of differentially labeled isocitric acid molecules from TCA metabolism of labeled oxaloacetic acid.[9]

in the Krebs formulation of the TCA cycle. To quote from the conclusions of Wood et al.

It is fairly certain that α-ketoglutarate is formed by combination of a 4-carbon dicarboxylic acid and pyruvate in this dissimilation. The exact mechanism of the reaction involved in this conversion is not known, however. This fact is emphasized by Krebs' erroneous prediction that fixed carbon would be found in both carboxyls of α-ketoglutarate in the case of pigeon liver dissimilation. Actually, the fixed carbon is confined to one carboxyl [Table I]. *This fact definitely precludes citrate as an intermediate,** since α-ketoglutaric acid derived from the symmetrical citrate molecule should contain equal amounts of fixed carbon in its two carboxyl groups.[10]

Similarly, Evans and Slotin concluded "the localization, however, of the entire radioactivity of the α-ketoglutarate in the carboxyl group adjacent to the carbonyl group as demonstrated in [their] Table II is *definite evidence against the suggested formation of this compound from citric acid.*"*[8] This state of

* My italics.

affairs persisted for several years, until Ogston, in a short note, pointed out that the experimental evidence had been misinterpreted and did not, in fact, preclude citrate as an intermediate.[11] Ogston stated "On the contrary, it is possible that an asymmetric enzyme which attacks a symmetrical compound can distinguish between its identical groups." The minimum conditions proposed for this discrimination between "identical groups" was a combination between the "symmetrical" substrate and the enzyme at three points (a', b', c'), with sites (a') and (c') of the enzyme being catalytically different. Ogston's proposal is illustrated in Figure I-5 for citric acid with a schematic three-point attachment.

In Figure I-5A we can see that the (a_1), (b), (c), functional groups on the citric acid substrate molecule interact with their respective binding sites (a'), (b'), (c'), on the enzyme surface. If we postulate that the conversion of citric acid to isocitric acid involves the cooperation of catalytic sites situated along the line determined on the enzyme surface by points (a') and (b'), then the hydroxyl function of the isocitric acid product will be found adjacent to that carboxyl function designated as (a_1) in the original citric acid molecule. The binding scheme of Figure I-5B illustrates the situation that would prevail if the citric acid substrate molecule were rotated so that the chemically like, paired grouping (a_2) binds to enzymatic site (a'). Binding of group (c) to its specific site (c') on the enzyme now results in the hydroxyl function (b) being directed *away* from its specific binding site (b') on the enzyme surface. Citric acid binding of the second type, therefore, does not lead to the proper

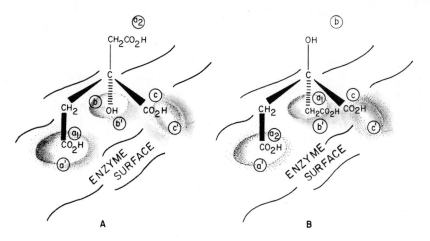

Figure I-5.   Schematic three-point substrate-enzyme binding.

positioning of the hydroxyl function of citric acid along the enzymatically active (a′—b′) site. Such binding would be nonproductive and could not result in the conversion of citric acid into isocitric acid. Since only binding of type A is catalytically active, the hydroxyl function in the isocitric acid product will be specifically located adjacent to that carboxyl designated as (a₁) in the original citric acid substrate (see above). Ogston's model involving attachment of a substrate molecule at three distinct sites on an enzyme surface can thus explain why only one of the two labeled isocitrate molecules anticipated by Krebs was experimentally observed. Ogston also noted that his model for biological stereospecificity between paired groups was directly related to the more familiar and historical manifestations of biological stereospecificity involved in the production of optical activity; "such combination is, of course, necessary whenever a single optical antipode is formed enzymatically from an inactive precursor." [11]

The differentiation of the paired carboxymethyl ($-CH_2CO_2H$) groupings of citric acid in the reactions of the TCA cycle has come to be known as the Ogston effect. As often occurs, the phenomenon is named after the scientist who supplied the key to proper interpretation, rather than after the experimentalist who initially observed the effect.

Although Ogston's insight suggested the correct interpretation of the labeling pattern observed during the operation of the TCA cycle, certain of his original statements were not precisely correct. Ogston's initial proposal stated that enzymes can distinguish between the "identical" groups of a "symmetrical" compound. Whatever Ogston's intended meaning, this statement led to an unfortunate amount of confusion among biochemists. The Ogston effect was interpreted to mean that identical groups on symmetrical molecules were distinguished by enzymes in some mysterious biological manner involving three points on an enzyme surface. The three-point-attachment concept was perverted to mean that the formation of citrate and its conversion to isocitrate occurred via totally enzyme bound intermediates; that is, the free "symmetrical" citrate molecule was never actually present in the reaction media. In actual fact, Potter and Heidelberger demonstrated in 1949 that biologically formed citrate that was isolated, purified, and then added to a second enzyme preparation still resulted in the formation of specifically labeled α-ketoglutarate.[12]

The misconceptions regarding this aspect of enzyme stereospecificity can largely be traced to an imprecise use of the term "symmetrical." A careful, critical examination of the representations in Figure I-5 should lead to the observation that the three-point attachment to the enzyme surface is *not inducing asymmetry* into the citric acid molecule; rather, the enzymatic attachment pictured just illustrates schematically how *an inherent lack of symmetry in the citric acid molecule may be recognized and exploited by an enzyme*. To reemphasize the point, citric acid should not be called a symmetrical molecule. It possesses some types of symmetry and lacks others. It is incorrect to say that a symmetrical citric acid molecule binds to an enzyme to form an asymmetric complex, which then permits differentiation between identical groups. Actually, because the citric acid molecule lacks certain elements of symmetry, the chemically like, paired carboxymethyl groups are *inherently nonequivalent*, and therefore it is possible to differentiate them.

The view that a "symmetrical" citrate molecule reacts via an "asymmetric" complex may, temporarily at least, lead to the feeling that the Ogston effect is understood. This incomplete view, however, cannot readily explain why the two carboxyl groups (and the two methylene carbons) of succinate are *not* capable of being differentiated and that metabolism through succinate will thus lead to randomization of isotopic carbon atoms introduced into the TCA cycle as specifically labeled acetate or oxaloacetate (cf. Figure I-3). A student whose "understanding" of the Ogston effect is only partially correct will also have difficulty in explaining why the succinate molecule contains two pair of enzymatically differentiable methylene hydrogens. These facets of succinate structure are depicted in Figure I-6.

The realization that Ogston's three-point-attachment model is actually a mechanistic device for demonstrating the inherent nonequivalance of chemically like, paired groups has led to analyses of the precise structural features of molecules necessary for manifestations of this type of enzymatic stereospecificity. Analyses of the structural elements present in molecules demonstrating this type of phenomenon (in both chemical and biochemical systems) have been carried out by Wilcox,[13] Hanson,[14] Hirschmann,[15] and Schwartz and Carter.[16] These authors differ in their approach, but all the analyses emphasize that a fundamental understanding of the metabolism of substrates such as citrate and succinate is dependent on an understanding of the relations between molecular symmetry and enzymatic stereospecificity.

For example, an analysis of the Ogston effect reveals that it is dependent on two biological stereospecific reactions—the formation of citrate catalyzed by citrate synthetase (Citrate oxaloacetate-lyase) and the formation of isocitrate catalyzed by aconitase [Citrate (isocitrate) hydro-lyase]. The citrate

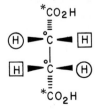

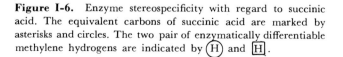

Figure I-6. Enzyme stereospecificity with regard to succinic acid. The equivalent carbons of succinic acid are marked by asterisks and circles. The two pair of enzymatically differentiable methylene hydrogens are indicated by Ⓗ and ⊞.

synthetase enzyme recognizes a particular *nonsymmetrical* feature of the oxaloacetate substrate molecule and forms a citrate product that is labeled differentially from acetate and oxaloacetate (Figure I-3). The aconitase enzyme, in turn, recognizes a particular *nonsymmetrical* feature of the citrate molecule and converts the differentially labeled citrate into a differentially labeled isocitrate product (Figure I-3). Neither of the enzymes is unique. Quite the contrary, each manifests only the biological stereospecificity normally observed with substrates possessing the symmetry properties of oxaloacetate and citrate.

We return to this phenomenon of biological stereospecificity between chemically like, paired groups in Chapters V and VI, after an elementary analysis of some aspects of molecular symmetry.

REFERENCES

1. T. S. Patterson and C. Buchanan, *Ann. Sci.*, **5**, 288 (1945).

2. "Researches on the Molecular Asymmetry of Natural Organic Products," L. Pasteur, *Alembic Club Reprints*, **14**, p. 28, University of Chicago Press, 1906.

3. *Ibid.*, p. 30.

4. "The Eightieth Anniversary of the Asymmetrical Carbon Atom," A. Sementsov, *Amer. Scientist*, **43**, 97 (1955). This article emphasizes the often overlooked differences between the proposals of Le Bel and van't Hoff.

5. T. M. Lowry, *Optical Rotatory Power*, p. 39, Longmans, Green and Co., 1935.

6. *Ibid.*, p. 42.

7. H. G. Wood, C. H. Werkman, A. Hemingway, and A. O. Nier, *J. Biol. Chem.*, **139**, 483 (1941).

8. E. A. Evans, Jr., and L. Slotin, *J. Biol. Chem.* **141**, 439 (1941).

9. H. A. Krebs, *Nature*, **147**, 560 (1941).

10. H. G. Wood, C. H. Werkman, A. Hemingway, and A. O. Nier, *J. Biol. Chem.*, **142**, 31 (1942).

11. A. G. Ogston, *Nature*, **162**, 963 (1948).

12. V. R. Potter and C. Heidelberger, *Nature*, **164**, 180 (1949).

13. P. E. Wilcox, *Nature*, **164**, 757 (1949).

14. K. R. Hanson, *J. Amer. Chem. Soc.*, **88**, 2731 (1966).

15. H. Hirschmann, *J. Biol. Chem.*, **235**, 2762 (1960); H. Hirschmann, in *Comprehensive Biochemistry*, Vol. 12, Chapter VII, M. Florkin and E. H. Stotz, Eds., Elsevier, 1964.

16. P. Schwartz and H. E. Carter, *Proc. Natl. Acad. Sci. (U.S.)*, **40**, 499 (1954).

Chapter II

INTRODUCTION TO MOLECULAR SYMMETRY

Unfortunately, the term symmetry is used by both scientist and nonscientist. Nearly everyone will recognize that a pine tree, a football, a triangle, and, yes, a citric acid molecule are "symmetrical" in an aesthetic sense (Figure II-1). However, difficulty in fully understanding those aspects of biological stereospecificity which are responsible for the Ogston effect is generally due to the *mistaken opinion that a citric acid molecule is "symmetrical" in a scientifically valid sense*. Even modern biochemistry texts frequently mislead the introductory student concerning this point. In a recently published text, for example, the description of the Ogston effect begins with the following statement:

Condensation of oxaloacetate with carboxyl-labeled acetate would be expected to produce citrate labeled in one carboxyl group [Figure 16-13], but since citrate is *a completely symmetrical molecule,** one would expect the two terminal carboxyl groups to be chemically indistinguishable.[1]

Then, following a description of Ogston's proposed three-site enzyme-substrate attachment, we find the additional statement "Succinate dehydrogenase shows no ability to discriminate between the two carboxyl groups of succinate, *presumably because it binds succinate through only two points of attachment."** [1] Such potentially misleading errors of fact are by no means confined to one text or a single author.

A recent edition of another popular biochemistry text states

In some enzyme-catalyzed reactions, the substrate is symmetrical from the point of view of organic chemistry. Glycerol and citric acid can be considered in this category since they have a plane of symmetry.

* My italics.

Figure II-1.   Some "symmetrical" objects.

It has been shown, however, that these compounds behave asymmetrically when serving as substrates for enzymes. *That is, $C_{a_1 a_2 bd}$, though symmetrical, is preferentially attacked at a_1 but not at a_2, although both groups are identical.*[2]

The remainder of this book is largely devoted to attempts to correct such erroneous views of biological stereospecificity between paired groups. As a first step, it is necessary to develop a valid understanding of the concept of molecular symmetry. An elementary comprehension of molecular symmetry, in turn, will permit us to clearly recognize geometrically nonequivalent paired groups (e.g., the carboxymethyl branches of citric acid) and to contrast such groups with equivalent paired groups (e.g., the carboxyl groups of succinate). Contrary to the statements quoted previously, only *geometrically nonequivalent paired chemical groupings of nonsymmetrical molecules are subject to biological stereospecific reaction. Equivalent chemical groups are never subject to stereospecific reaction*—even in biological systems—no matter how many binding sites are invoked.

All the objects in Figure II-1 possess some symmetry. Yet it is also true that none of these objects possess the symmetry of a perfect sphere, and some are "more symmetrical" than others. What is needed is a scientific definition of symmetry that will permit us to describe precisely the different types of symmetry that are present in the objects of Figure II-1, as well as how each differs from the symmetry of a sphere. Such precise scientific descriptions of the symmetries of molecules will, of course, be directly related to the inherent molecular geometries. Since enzymatic stereospecificity between the chemically like, paired groups of a given molecule, or between two enantiomeric molecules, is based on molecular geometries, we will be able to relate the scientifically defined molecular symmetries of substrates to different manifestations of biological stereospecificity. The situation is analogous to understanding the enzymatic reaction specificity for various precisely defined chemical functional groups on molecules in terms of the different chemical reactivities taught in beginning organic chemistry. One type of biological specificity is based on three-dimensional geometry; the other on chemical functionality.

* My italics.

SYMMETRY OPERATIONS AND SYMMETRY ELEMENTS

The scientific definition of symmetry begins with a description of two intimately related terms, *symmetry operations* and *symmetry elements*. Both these terms are necessary in defining symmetry, and they are so inextricably linked that they lack independent meaning. The terms refer, however, to different *kinds* of things. As the name implies, a symmetry operation is an operation, a precisely programmed movement of an object. It is a movement of an object that is performed, either mentally or by the use of models, such that the final orientation is *equivalent* (indistinguishable but not necessarily identical) to the initial orientation. Let us use the isosceles triangle in Figure II-2 as a representative object.

The geometry of this triangle is such that it can be moved (i.e., turned over) so that the unique vertex labeled 2 of the initial *conformation* (three-dimensional orientation of the object) lies on 2 of the final conformation, whereas the two equivalent vertices labeled 1 and 3 are interchanged. The two conformations are called *equivalent* because someone who had not observed the operation could not distinguish the final representation from the initial representation. A nonobserver would therefore be unable to ascertain whether any operation had actually been performed. (The linear translation of the triangle from the left side of the page to the right side is only for diagrammatic representation and is not part of the symmetry operation. The symmetry operation of rotation described would actually result in two superimposed triangles with 1 on 3, 2 on 2, and 3 on 1.) The operation of "flipping" the triangle over thus constitutes a symmetry operation. The two representations (before and after the operation) are *equivalent*, but they are not *identical*, since, as demonstrated, if the vertices of the triangle were actually numbered, it would be observed the equivalent vertices 1 and 3 are interchanged. Equivalence of the initial and final conformations defines a symmetry operation. Absolute identity of the conformations is not required.*

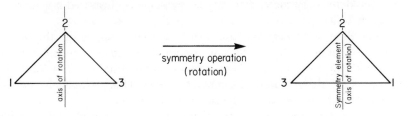

Figure II-2. Symmetry operation and element of an isosceles triangle.

* Misunderstanding of this important point is at least partially responsible for the incorrect view that the paired carboxymethyl functions of citric acid are identical. The "symmetry" of the citric acid molecule is due to a molecular mirror plane. The symmetry operation of reflection through this plane does interchange the two paired carboxymethyl

The symmetry operation described previously is a precisely defined movement because we specified that vertex 2 remains fixed and vertices 1 and 3 are interchanged. Thus this symmetry operation is a rotation of the triangle, defined with respect to an axis of rotation which lies in the plane of the triangle and bisects the unique angle 2. A *symmetry element* is a geometrical entity such as a line, a plane, or a point with respect to which one or more symmetry operations are defined and may be carried out. In the foregoing example, the line is a symmetry element, an axis of symmetry, with respect to which the symmetry operation of rotation is applied.

The isosceles triangle of Figure II-2, possesses additional symmetry elements. Because the structure is planar, the plane of the paper is a symmetry plane. Reflection through this plane (with the bottom of the planar figure becoming the top and the top becoming the bottom) obviously results in an equivalent orientation, and this reflection qualifies as a second symmetry operation. Perhaps the symmetry operation of reflection can be visualized more clearly with respect to the third symmetry element of the isosceles triangle, the vertical symmetry plane. This symmetry plane is perpendicular to the plane of the triangle and bisects the unique angle 2. The intersection of this plane of symmetry with the plane of the triangle is the rotational axis symmetry element defined previously. The symmetry operation of reflection can be visualized if we imagine this vertical symmetry plane to be a double mirror surface. The reflection of the left side of the triangle in this mirror plane becomes the right side of the final conformation, and the reflection of the right side of the original representation in the mirror plane becomes the left side of the final conformation. This symmetry operation leaves point 2 fixed, since it lies on the symmetry plane, and interchanges vertices 1 and 3. The result of this movement by reflection is thus a conformation *identical* to that obtained by the rotational symmetry operation discussed previously, and, of course, it is *equivalent* to the initial conformation. The operation of reflection defined by the vertical symmetry plane is illustrated schematically in Figure II-3.

The three symmetry operations that have been applied to the isosceles triangle of Figure II-3 are the only unique movements applicable to this object that will result in equivalent initial and final conformations. Therefore, the associated three symmetry elements (Figure II-4) completely and precisely define the symmetry of the isosceles triangle.

Despite the very large numbers of different chemical molecules that exist, it has been found that there are comparatively few combinations of

---

functions of the molecule to create an *equivalent conformation.* Chemically paired groups that are interchanged *only by reflective symmetry operations,* however, are *not geometrically equivalent* in an absolute sense. (Cf. Table II-3 and the related discussion.)

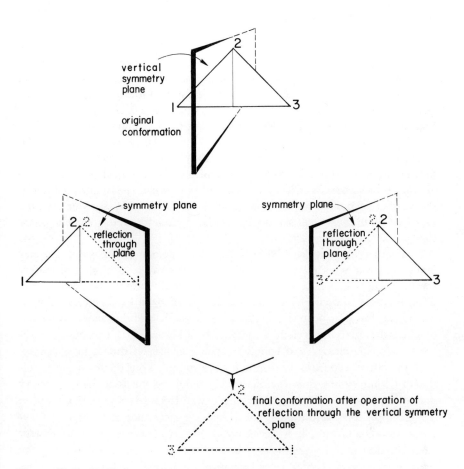

Figure II-3. Reflection through the vertical symmetry plane of an isosceles triangle.

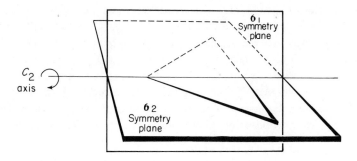

Figure II-4. Symmetry elements of the isosceles triangle.

symmetry elements that can occur in these molecules. These permissible combinations of symmetry elements may be grouped together into a total of 32 *symmetry point groups* (possible combinations of symmetry operations that leave a specific point of the molecule unchanged). Objects such as an isosceles triangle or a water molecule possessing the symmetry elements of Figure II-4 may be described as members of the C_{2v} point group.

In this book we are not concerned with the classifications of molecules or objects into their respective point groups. For students who are interested in learning more about this aspect of chemistry and chemical structure, brief articles in the *Journal of Chemical Education*[3,4] provide a ready introduction to the subject of point groups. Additional information may be found in the excellent published treatises on molecular symmetry and stereochemistry that are listed on page 293.

Whether an isosceles triangle is stated to belong to the C_{2v} symmetry point group or whether the symmetry elements of the triangle are merely summarized as in Figure II-4, the final result is the same. The "symmetry" of an isosceles triangle, which was intuitively realized by merely looking at the object, has now been defined in a scientifically useful manner. If we applied similar analyses to each of the objects of Figure II-1, it would be possible to describe clearly and concisely how each object differs in symmetry from the others and from the symmetry of a sphere. Thus the scientific definition of symmetry in terms of symmetry elements and the associated symmetry operations performs the initial task outlined in the beginning of this chapter. To accomplish the additional goal of relating molecular symmetry to enzymatic stereospecificity, it is necessary to define all the types of symmetry elements that may be present within a molecule and to learn to recognize their presence or absence in given biological substrates. Table II-1 summarizes the possible symmetry elements, the associated symmetry operations, and the symmetry notations. *

Inspection of Table II-1 illustrates that the task of scientifically defining the symmetry of a molecule is quite simple. Only four different types of symmetry elements are required; two which were introduced in the discussion of the isosceles triangle. By confining our goal to understanding biological discrimination between chemically like, paired groups and between enantiomeric molecules, it will be possible to introduce further simplifications. Following a discussion of each set of symmetry operations and elements, we examine their application to the problem of biological stereospecificity.

---

* The notation used throughout this book is that commonly used by spectroscopists and is called the Schoenflies notation. Crystallographers use somewhat different notations called the Hermann-Marquin terms.

Table II-1   Symmetry Elements, Notations, and Operations

Element	Notation	Operation
1. Proper rotation axis	C_n	Rotation about an axis by $360/n$ degrees
2. Center of symmetry (inversion center)	i	Inversion of all points through inversion center
3. Plane of symmetry (mirror plane)	σ	Reflection in the plane
4. Improper rotation axis	S_n	Rotation about an axis by $360/n$ degrees and reflection in a plane perpendicular to the axis

THE PROPER ROTATION AXIS, C_n

A molecule possesses a proper axis of rotation of order n if a rotation about the axis of $(360/n)°$ produces a new conformation indistinguishable from the original conformation. In the discussion of the symmetry of the isosceles triangle we saw that a rotation of 180° was required to produce an equivalent conformation; thus the appropriate symmetry element is indicated in Figure II-4 as C_2 $[(360/2)° = 180°]$.

All molecules possess an infinite number of trivial C_1 axes because rotation by 360° about any axis restores the molecule to its original orientation. At the other extreme, linear molecules such as acetylene or nitrogen possess a C_∞ axis coincident with the linear molecular axis, since rotation by even the slightest amount, $(360/\infty)°$, about this molecular axis results in a new equivalent orientation. The linear acetylene or nitrogen molecule possesses two additional proper axes of rotation, mutually perpendicular C_2 axes passing through the midpoint of the molecule as indicated in Figure II-5.

Chloroform offers an example of a molecule having a C_3 axis. Rotation about the axis defined by the unique carbon–hydrogen bond leaves the carbon and hydrogen atom positions unchanged. As illustrated in the planar projection, the three chlorine atoms differ in position by 120°. Therefore, a rotation about the carbon–hydrogen bond axis of $(360/3)°$ will rotate the chlorine atoms into new equivalent positions.

Benzene has an easily recognizable C_6 proper axis of rotation perpendicular to the plane of the molecule and passing through the geometric center. The reader should also be able to visualize the six additional C_2 axes, lying in the molecular plane. Three of these bisect the carbon–carbon bonds and the other three bisect the carbon–carbon–carbon angles (Figure II-5).

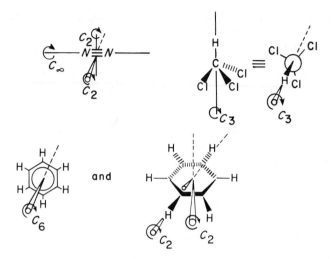

Figure II-5.   Examples of molecules possessing C_n symmetry elements.

We leave it to the reader to convince himself by studying models or perspective drawings that the tetrahedral methane molecule possesses four, three-fold proper axes, and three, two-fold axes.

CENTER OF SYMMETRY, i

A molecule possesses a center of symmetry or inversion center if a straight line can be drawn from each atom (the symmetry operation) which passes through the center of the molecule (the associated symmetry element, i); then, continuing in the same direction, the line encounters an equivalent atom equidistant from the center. In other words, if there is a point within a molecule that can be assigned the coordinates 0, 0, 0, such that changing the coordinates (x, y, z) of all atoms to $(-x, -y, -z)$ results in an equivalent configuration, then the molecule possesses a center of symmetry at 0, 0, 0. A center of symmetry is depicted in Figure II-6 for *trans*-1,3-dimethylcyclobutane.

The points of intersection of the rotational axes in the nitrogen and benzene molecules in Figure II-5 are also centers of symmetry.

It should be noted that, since the inversion operation reflects each atom through the center into an equivalent atom, with the exception of one unique atom which may lie on the center, the atoms of a molecule possessing an inversion center must occur in pairs located equidistant and in opposite

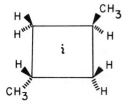

Figure II-6. Representative inversion center for *trans*-1,3-dimethylcyclobutane.

directions from the center. The tetrahedral methane molecule does *not* possess an inversion center, since the paired hydrogen atoms are *not* located in opposite directions from the unique center carbon atom.

PLANE OF SYMMETRY, σ

A plane of symmetry is a plane that passes *through the molecule* such that reflection of all atoms through the plane transforms the molecule into an equivalent conformation. The operation of reflection performed with respect to a symmetry plane σ was illustrated in Figure II-3. Linear molecules such as the nitrogen molecule in Figure II-5 possess an infinite number of such symmetry planes which intersect each other along the molecular axis. All planar molecules, such as benzene in Figure II-5, have at least one symmetry plane that coincides with the molecular plane. The higher symmetry of benzene, however, results in the presence of six additional planes of symmetry. These are perpendicular to the molecular plane and intersect the molecular plane along the six C_2 axes of rotation described previously.

The tetrahedral dichloromethane molecule (CH_2Cl_2) is another good example of a molecular structure characterized by symmetry planes. The CH_2Cl_2 molecule (Figure II-7) possesses two mutually perpendicular symmetry planes. One plane contains the carbon and two hydrogen atoms, and

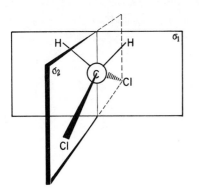

Figure II-7. Symmetry plans of dichloromethane.

the second plane is determined by the carbon and chlorine atoms. The latter example emphasizes that, if a molecule contains only a single representative of a type of atom or group (the carbon atom in CH_2Cl_2), this unique atom or group must lie on all symmetry planes of the molecule.

An analysis of the more symmetrical tetrahedral methane molecule should lead to the observation that six planes of symmetry are present.

THE IMPROPER ROTATION AXIS, S_n

The final symmetry operation to be examined is defined by the improper rotation axis, called S_n. The operation defined by this element consists of two manipulations; rotation about an axis by $(360/n)°$, followed by reflection through a plane *passing through the molecule, perpendicular to this S_n axis*. These two manipulations have been described separately as the symmetry operations of rotation about an axis and reflection through a plane, defined by the elements C_n and σ, respectively. The combination of the two operations and the defined relationship of the axis and the plane, however, form a distinctly different symmetry operation. An S_n symmetry element may therefore exist when neither the C_n nor the perpendicular σ elements exist separately. Consider, for example, the allene molecule in Figure II-8.

Note that the allene molecule is not planar, but has the two hydrogen–carbon–hydrogen groups oriented at a 90° angle. This is most readily seen in the end-view perspective. First observe that the linear carbon–carbon–carbon molecular axis defines a C_2 proper axis of rotation, since a 180° rotation produces an equivalent configuration. A C_4 element does not exist, however, since a 90° rotation produces an obviously altered orientation. A σ element passing through the central carbon atom perpendicular to the molecular axis is also nonexistent, since this reflection would interchange the front and back hydrogen–carbon–hydrogen groups, and these clearly have different absolute orientations. However, an S_4 element defining the above-mentioned sequence of rotation and reflection does exist. This is diagrammed in Figure II-9.

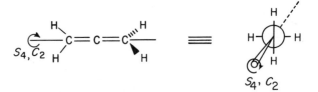

Figure II-8. S_4 element of allene.

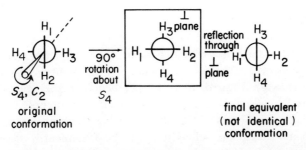

Figure II-9. Demonstration of the symmetry operation defined by the S_4 axis of allene. *
* The reader should be able to place the additional symmetry elements of allene: Two mutually perpendicular C_2 axes passing through the central carbon atom and bisecting the carbon hydrogen dihedral angles (see Figure II-11 for a definition of dihedral angles) and two mutually perpendicular planes of symmetry, each containing one of the hydrogen–carbon–hydrogen groupings and the molecular carbon–carbon–carbon axis.

The tetrahedral methane molecule offers a related example of S_4 symmetry. Earlier we indicated that the methane molecule possesses three C_2 axes. These axes are mutually perpendicular and each is also an S_4 axis (Figure II-10). Note that although the C_2 and S_4 axes happen to coincide in

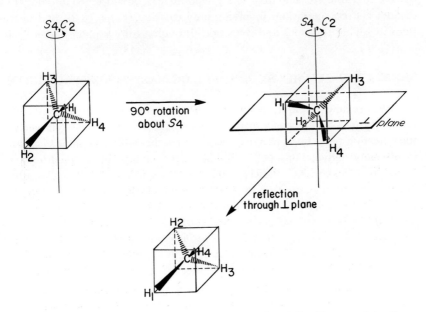

Figure II-10. Demonstration of the symmetry operation defined by one of the S_4 axes of methane.

these cases, neither C_4 axes nor perpendicular σ elements exist as independent symmetry elements.

Since a 360° rotation always leads to an identical conformation, the C_1 proper rotational axis is a trivial element. The same is not true, however, of the S_1 improper rotational axis. Rotation by 360° about an axis followed by a reflection through a mirror plane perpendicular to the axis (the operation defined by S_1) will result in an equivalent conformation only in special cases. In particular, the S_1 axis will exist only in molecules possessing a σ element (the perpendicular plane defined by S_1). The S_1 element is therefore equivalent to a σ symmetry plane. In the same way, the S_2 element may be seen to be equivalent to the i inversion center. Rotation about an axis by 180° followed by reflection through a plane perpendicular to this axis will transform the coordinates of an atom from x, y, z, to $-x$, $-y$, $-z$, just as the operation defined by i. (The 0, 0, 0, coordinates of the center of symmetry will be the point of intersection of the S_2 axis with the perpendicular mirror plane.) Because of established thought patterns, the existence of the elements of symmetry σ and/or i is frequently easier to discern in a molecule than is the existence of the equivalent S_1 and/or S_2 elements. However, the realization that the symmetry operations defined by σ and i can also be defined by S_1 and S_2 leads directly to the observation that the four symmetry operations defined in Table II-1 and discussed previously can be simplified into only two distinct classes; operations involving only rotation (C_n, $n \geq 2$), and operations involving rotations and reflection through a mirror plane (S_n, $n \geq 1$).

MOLECULAR SYMMETRY AND BIOLOGICAL STEREOSPECIFICITY

We are now prepared to relate enzymatic stereospecificity and molecular symmetry in a formal and precise manner. If a molecule lacks symmetry of the reflection type (no S_n axis, $n \geq 1$), the structure will be capable of existing in optically active, nonsuperimposable mirror image forms (enantiomers). A wealth of experience indicates that enzymes will generally differentiate between enantiomers, either by reaction with one of the two enantiomers as a substrate or by creation of one enantiomer as a product. On the other hand, *if a molecule lacks symmetry or the rotational type (no C_n, $n \geq 2$), the chemically like, paired groups of the molecule must be nonequivalent.* Even in molecules possessing elements of rotational symmetry, it is possible that the chemically like, paired groups will not be interchanged* by any of the rotational sym-

---

* By "interchanged" we mean that, when the equivalent conformation resulting from a symmetry operation is superimposed upon the original conformation, like group 1 of the

metry operations. *Such paired groups are also nonequivalent.* Studies of enzymatic reactions, mainly through the use of carbon and hydrogen isotopes, have shown that enzymes will differentiate between the chemically like, but non-equivalent groups of both classes of molecules. Thus the lack of rotational symmetry elements in a molecule that will interchange the chemically like, paired groups (*rotational asymmetry*) will result in biological stereospecificity between these nonequivalent paired groups of a particular substrate molecule. On the other hand, the lack of reflective symmetry elements in a molecule (reflective asymmetry or *dissymmetry*) will result in enzymatic stereospecificity between separate enantiomeric substrate molecules. The analysis of the two aspects of biological stereospecificity in terms of these contrasting symmetry properties of the substrate molecules is originally due to Hirschmann, who also pointed out that both aspects of enzyme stereospecificity are due funda-mentally to the fixed alignment of a substrate to a dissymmetric three-dimensional catalytic site on an enzyme. In Hirschmann's words

The essential *difference* between the two manifestations of enzymatic stereospecifici-ty does not lie in the role of the enzyme but solely in that of the substrate. Discrimina-tion between optical antipodes depends upon their reflective asymmetry, whereas the discrimination between like substituents depends on the rotational asymmetry of the substrate with respect to these substituents.[5]

The relations between some possible categories of substrate symmetry and different manifestations of enzymatic stereospecificity are summarized in Table II-2 along with some useful nomenclature. It is also instructive to emphasize the symmetry relations existing between various chemically like, paired groups within individual molecules. The potential relationships be-tween paired groups are summarized in Table II-3.

As indicated in Table II-3, there are three possible classifications of chemically like, paired groups. Chemically like, paired groups that are inter-changed by rotational symmetry operations are called *equivalent*. The sepa-rate groups comprising equivalent pairs are distinguishable only in theory, through the use of arbitrary and artificial labels, for example. Equivalent paired groups, therefore, *cannot be differentiated in an actual chemical reaction—even an enzymatically catalyzed chemical reaction.* The successive replacement of the two equivalent paired groups by a nondissymmetric test group (e.g., a deuterium atom) does not produce separate, isomers but equivalent substi-

---

equivalent conformation is superimposed upon like group 2 of the original conformation and group 2 of the equivalent conformation is superimposed upon group 1 of the original conformation.

Table II-2   Some Possible Relationships Between Substrate Symmetry and Enzyme Stereospecificity

Possible substrate symmetry elements	Symmetry designation	Empirical observation	Type of enzyme stereospecificity anticipated
Possesses rotational symmetry but lacks reflective symmetry, i.e., no σ, i or S_n ($n \geq 3$).	*Dissymetric* substrate	Exists in optically active nonsuperimposable mirror image forms. Chemically like, paired groups *may be* superimposable by rotation and thus equivalent.	Enzymatic reactions with, or formation of, one of the isomeric enantiomers. Enzymatic stereospecificity between the equivalent paired groups is impossible.
Possesses reflective symmetry but lacks rotational symmetry, i.e., no C_n ($n \geq 2$).	*Nondissymmetric* substrate	Optically inactive; but chemically like, paired groups will be nonequivalent.	Enzymatic reactions specifically with one of the chemically like, paired groups of an optically inactive substrate molecule.
Lacks both reflective and rotational symmetry, i.e., possesses only C_1.	*Asymmetric* substrate	Optically active. Similar groups, including chemically like, paired groups, will be chemically and geometrically distinct.	Enzymatic reactions with one enantiomer, plus usual enzymatic specificity for distinctly different chemical groups.

tuted molecules (superimposable). It can be seen from the information summarized in Table II-2 that only dissymmetric molecules may possess equivalent paired groups.

Two classes of chemically like, nonequivalent, paired groups can be designated which *are* subject to differentiation in enzymatically catalyzed chemical reactions. If the chemically like, paired groups of a molecule are interchanged only by reflective symmetry operations, they are called *enantiotopic* (*topos* = place).[6] The separate groups comprising enantiotopic pairs may be differentiated by a dissymmetric reagent or in a dissymmetric environment. Enzymes, of course, are dissymmetric reagents and will differentiate between enantiotopically related groups. As indicated in Table II-3,

Table II-3   Possible Symmetry Relationships Between the Chemically Like, Paired Groups Within Individual Substrate Molecules

Designation	Symmetry relationship	Chemical substitution result (by nondissymmetric test group)	Conditions for differentiation
Equivalent groups	Groups interchanged by C_n symmetry operation ($n > 1$).	No isomers generated—equivalent molecules.	Only by artificial devices; impossible experimentally.
Enantiotopic groups	Groups interchanged by S_n symmetry operation ($n \geq 1$).	Enantiomeric stereoisomers generated.	Only by a dissymmetric reagent or environment. Enzymes!
Diastereotopic groups	Groups are not interchanged by any symmetry operation.	Diastereomeric stereoisomers generated.	By an reagent or environment of sufficient physicochemical specificity. Enzymes!

the successive replacement of enantiotopic paired groups with a nondissymmetric substituting test group will lead to the production of stereoisomeric molecules that are enantiomeric. From Table II-2 it can be seen that only nondissymmetric molecules can possess enantiotopic paired groups.

Finally, we see that chemically like, paired groups of a molecule may be classified as *diastereotopic*. The separate groups comprising a diastereotopic pair may be differentiated by physicochemically selective reagents or instruments. Enzymes function effectively as such selective reagents and will readily differentiate between diastereotopically related groups. The successive replacement of diastereotopic paired groups by a substituting test group will lead to the production of distinct stereoisomeric molecules that are *not* related as object and nonsuperimposable mirror image. Such stereoisomers are termed *diastereomers*. Chemically like, paired groups in asymmetric substrates must be diastereotopic. Since, however, the symmetry axes of molecules may well be situated so that some of the paired groups are not interchanged by the various symmetry operations, diastereotopic paired groups are not confined to asymmetric molecules, but may be found in any of the three classes of substrates listed in Table II-2.

It should be mentioned that chemically like groups are not necessarily confined to *paired groups*. There may be more than two equivalent, enantio-topic or diastereotopic groups or atoms within a given molecule. For example, allene contains four equivalent hydrogen atoms (cf. Figure V-12), and the three methyl hydrogens of acetic acid are also equivalent (cf. Chapter VI, pp. 193–211). Yet most molecules of biochemical interest possess a maximum of two chemically like groupings, and for this reason, we will continue to refer to chemically like, paired groups.*

CONFORMATIONAL ISOMERS AND MOLECULAR SYMMETRY

The relationships between molecular symmetry, chemically like, paired groups, and enzymatic stereospecificity (summarized in Tables II-2 and II-3) are analyzed in greater detail in Chapters III and V. Before initiating a more detailed analysis of these relationships, however, we must consider another aspect of molecular structure and its relationship to molecular symmetry.

In the preceding discussion, a symmetry operation was defined as the movement of an object that resulted in a final orientation or conformation equivalent to the initial orientation or conformation. In terms of objects such as the isosceles triangle or *rigid* molecular models, this description is quite satisfactory. However, rotation about carbon–carbon single bonds can be regarded as relatively "free." The required activation energies for these rotations are generally less than the approximately 20 kcal/mole barrier that is required to prevent such rotation at room temperatures due to the available thermal energy. This means, of course, that organic molecules are not rigid molecular structures but actually exist in a variety of interconvertible three-dimensional orientations or *conformers*. Three of the distinct *conformational isomers* of the ethane molecule are pictured in Figure II-11. These distinctly different geometrical groupings of C_2H_6 are readily defined in terms of the *dihedral angle* ϕ. Since these conformational isomers are interconverted by rotation about the carbon–carbon single bond, and since the energy maximum encountered during this rotation (in the eclipsed conformation) is only about 3 kcal/mole, the ethane molecule will exist as a dynamic mixture of such conformational isomers. The projections of Figure II-11 make it apparent that each of these conformational isomers possesses a different set of symmetry elements. In discussing the symmetry of an organic molecule such as ethane, therefore, we are obliged to specify which specific conformational isomer we are treating. The symmetry operation is then defined as a movement performed with respect to a certain element (e.g., the S_6 axis of the

---

* For a more detailed analysis of the possible stereoisomeric relationships of groups within molecules see the definitive chapter by K. Mislow and M. Raban, in *Topics in Stereochemistry*, Vol. 1, N. L. Allinger and E. L. Eliel, Eds., Wiley, 1967.

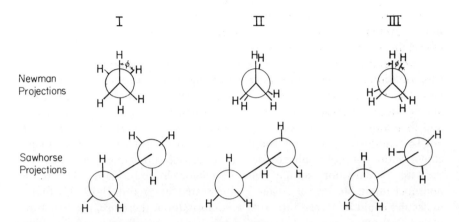

Figure II-11. Projections of some conformational isomers of ethane: I, staggered ($\phi = 60°$); II, eclipsed ($\phi = 0°$); III, skewed ($\phi < 60°$). Newman projections are projections onto the pages of molecules viewed along a carbon–carbon bond. The three intersecting lines represent the "front" or nearer carbon atom and three of its bonds; the circle represents the "back" or further carbon atom and its remaining three bonds.

staggered conformer or the C_3 axis of the eclipsed conformer) which transforms an artificially rigid or "frozen" conformational isomer from one orientation into an equivalent orientation. Although initially it might be felt that the existence of interconvertible conformational isomers is an additional complication, for many of our purposes, the existence of "free" rotations about carbon–carbon single bonds will be a simplifying factor. For example, in Figure II-6 we indicated that there is a center of symmetry i. The existence of this element requires the two methyl groups to be interchanged by reflection through the molecular center. If we were to consider the various conformational isomers that would be possible if "free" rotation about the CH_3—C bonds were not allowed, each of the six hydrogens attached to the methyl carbons could be "frozen" into unique geometric positions. Only a few of these possible conformations would then contain a center of symmetry where atoms H_a, H_b, H_c on one methyl group would be reflected into equivalent atoms H_a', H_b', H_c' on the opposed methyl group. However, under conditions where relatively free rotation about the CH_3—C bonds exists, the three methyl hydrogens become equivalent and, on a time-averaging basis, will sweep out a torus. The time-averaged positions of the hydrogens on the opposed methyl groups (opposing torus surfaces) are readily interchanged by reflection through the center of symmetry. In this case, the existence of ready conformational interconversions increases the symmetry possessed by the time-averaged molecule and aids in the placement of the appropriate symmetry elements. We will find that a consideration of the facile interconversions of cyclohexane conformational isomers also leads to a simplified analysis of certain stereochemical problems.

Recognizing that organic molecules usually exist as rapidly equilibrating conformational isomers necessitates modifying slightly the rules summarized in Tables II-2 and II-3 governing the relation between molecular symmetry and enzymatic stereospecificity. For a chemical compound to be dissymmetric and therefore capable of existing in enantiomeric forms, the molecule must not be readily (i.e., by "free rotation") able to assume *either* mirror image conformations *or* any single conformation possessing reflective symmetry. In the case of ethane, for example, the skew conformation in Figure II-11 has a nonsuperimposable mirror image as illustrated in Figure II-12. If ethane molecules could be frozen into a skew conformation, therefore, the structure would be dissymmetric and capable of existing in optically active forms. The Newman projections of Figure II-12 illustrate, however, that the enantiomeric skew conformations of ethane can be interconverted by permitted "free rotations" about the carbon–carbon single bond. For every dissymmetric ethane molecule represented by the left-hand projection of Figure II-12, there will be an ethane molecule of equivalent energy possessing the dissymmetric conformation represented by the right-hand projection. The dissymmetric contributions of these two skew ethane populations will therefore cancel and no net optical activity will be observed.

In the case of the ethane molecule, the foregoing analysis can be simplified by observing that both the staggered conformation and the eclipsed conformation of Figure II-11 contain elements of reflective symmetry (i and σ, respectively). The observation that such nondissymmetric conformations are readily accessible to the ethane molecule by "free" rotations precludes the possibility of ethane existing in distinct enantiomeric, optically active forms, *even though at any given moment most of the ethane molecules will exist in various skew conformations which lack reflective symmetry.*

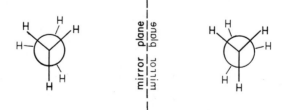

Figure II-12.   Enantiomeric skew ethane conformations.

In an analogous fashion, if a molecule written in the "most symmetrical" conformation that can be readily attained possesses rotational symmetry elements that interchange the chemically like, paired groups, then these paired groups will be equivalent, and enzymatic stereoselectivity between them will be impossible. This will be true, just as in the case of the possible optical activity of ethane discussed previously, even if the majority of the molecules at any moment exist in conformations that place the chemically like, paired groups in differing geometrical positions.

The molecule pictured in Figure II-13 demonstrates that certain unusual dissymmetric structures may be converted into their nonsuperimposable mirror image conformations by rotation about carbon–carbon bonds, even though, in contrast to ethane, they do not possess any accessible conformation with reflective symmetry.

The structure represented in Figure II-13 is an example of an overcrowded biphenyl system in which the presence of four large blocking *ortho* substituents (X) on the biphenyl system prevents rotation about the carbon–carbon bond joining the two phenyl rings and leads to a fixed perpendicular orientation of these two rings. Rotations about the carbon–carbon bonds joining the asymmetric (C_{abc}) *para* substituents to the phenyl rings, however, are permitted. The original projection is dissymmetric and not superimposable on its mirror image. As illustrated, however, a 90° rotation of both of the *para*-C_{abc} groups in the same direction results in a conformation that is superimposable on the original mirror image by a second 90° rotation of the whole molecule. Thus this dissymmetric structure exists as a mixture of enantiomeric pairs, which rapidly interconvert by free rotation about carbon–carbon single bonds. The substance is therefore optically inactive. Thus we

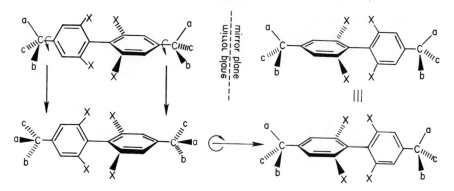

Figure II-13. Interconversion of nonsuperimposable mirror image conformers through dissymmetric intermediate states.

see that *racemization* (loss of optical activity) may sometimes proceed through intermediates that lack reflective symmetry.

In considering whether enzymatic stereospecificity between chemically like, paired groups can exist, we are concerned with the possibility of interchanging chemical groups by rotational symmetry operations. This interchanging of equivalent chemical groups requires the superpositioning of molecular forms (in contrast to molecular mirror image forms) by rotations. Since superpositioning of molecular forms of models cannot arise *unless* the molecule can be written in conformations that possess rotational symmetry, no situation analogous to the racemization pathway just described is possible. In examining the potential for enzymatic stereospecificity between paired groups of a substrate, therefore, we need only test for rotational symmetry and interchangeable equivalent groups in the conformations that have the greatest likelihood of possessing such symmetry. The separate chemical entities comprising all nonequivalent paired groups (enantiotopic or diastereotopic), whether present in molecules totally lacking rotational symmetry or in molecules lacking appropriately placed rotational axes, will be differentiated in enzymatically catalyzed processes (Table II-3).

As a final example of the complications and/or simplifications introduced into stereochemical analyses by considering the conformational changes permitted by "free" rotations about single bonds, we examine the case of cyclohexane derivatives. Cyclohexanes tend to exist in nonplanar, strain-free, puckered conformations referred to as chair conformations. By relatively "free" rotations about the carbon–carbon single-bonded framework (a rotational barrier of about 11 kcal/mole must be overcome), the chair conformations may be interconverted. In the case of a monosubstituted cyclohexane, this chair–chair interconversion transforms an axially bonded substituent into an equitorially bonded substituent. This interconversion is illustrated in Figure II-14. Due to steric interactions with the other axial

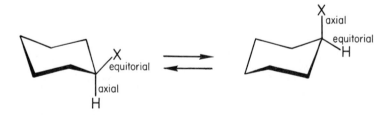

Figure II-14.   Chair–chair conformational interconversions of monosubstituted cyclohexanes.

bonds on the same side of the cyclohexane ring at positions 3 and 5, the conformation with the equitorial substituent is generally of lower energy and thus favored.* Since both of these conformations of monosubstituted cyclohexanes contain a σ plane perpendicular to the ring and passing through C-1 and C-4, neither conformation is dissymmetric and optical activity is not possible.

In the case of a disubstituted cyclohexane, however, the situation becomes more complex. With a 1,2-disubstituted arrangement, for example, the two substituting groups may be either *cis* or *trans*; and as illustrated in Figure II-15, each of these isomeric possibilities will also be found in two interconvertible chair conformations.

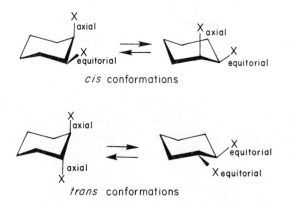

Figure II-15. 1,2-Disubstituted cyclohexane conformational isomers.

As Figure II-16 reveals, the two chair conformers of the *cis* isomer possess nonsuperimposable mirror image relationships. The *cis* disubstituted cyclohexane isomer thus falls into the same category as skew ethane and the biphenyl molecule of Figure II-13 in that *the potential enantiomeric forms are actually interconvertible conformational isomers.* The *cis*-1,2-disubstituted cyclohexane structures therefore will not be optically active.

In the case of the *trans*-1,2-disubstituted isomer, both chair conformations have nonsuperimposable mirror images. Furthermore, since the relative geometry of axial substitution is distinctly different from that of equitorial substitution, only a moment's reflection should lead to the realization that no mirror image relationship can possibly exist which would superimpose the diaxial substituents of one conformation onto the diequitorial substituents of

---

* Any modern organic chemistry text can be consulted for a more detailed review of these aspects of cycloalkane stereochemistry. See, for example, J. D. Roberts and M. C. Caserio, *Basic Principles of Organic Chemistry*, W. A. Benjamin, 1965, Chapter 4.

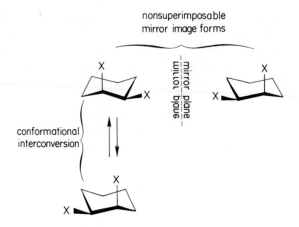

Figure II-16. Demonstration of the enantiomeric relationship of *cis*-1,2,disubstituted cyclohexane chair conformers.

the alternative chair conformation. The *trans* structure, therefore, will exist in optically active enantiomeric forms despite conformational flexibility.

On the other hand, the diaxial or diequitorial substitution pattern present in the *trans* stereoisomeric forms in Figure II-17 makes it possible to place

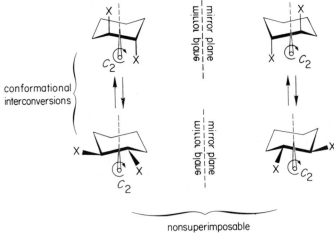

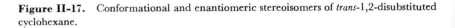

Figure II-17. Conformational and enantiomeric stereoisomers of *trans*-1,2-disubstituted cyclohexane.

C_2 axes bisecting the C-1–C-2 and C-4–C-5 bonds in each of these structures. The rotational symmetry operations then cause the interchange of the *trans* substituting groups on C-1 and C-2 (X_1 and X_2). The *trans*-1,2-disubstituted cyclohexanes, therefore, belong to a class of substrates possessing rotational symmetry elements that interchange the chemically like, paired groups. Such rotationally interchangeable groups are equivalent paired groups, and biological stereospecificity between the X_1 and X_2 groups is impossible. In contrast, note the substituents on the *cis* isomers of Figure II-16 have axial-equitorial substitution patterns. It should be clear that no rotational operation can cause the superpositioning of a 1-axial group onto a 2-equitorial group. The *cis* conformational isomers thus lack rotational symmetry and belong to a class of molecules in which biological stereospecificity between chemically like, paired groups would be permitted, despite conformational flexibility.

These same *stereochemical conclusions* regarding cyclohexanes can be readily derived if we totally ignore the actual conformational cyclohexane isomers present and view the cyclohexane ring as planar, and the substituents bonded in a perpendicular manner either above or below the plane of the ring. The *cis* and *trans*-1,2-disubstituted cyclohexanes would then be represented by the projections in Figure II-18. From these projections we can immediately conclude that the *trans* isomer possesses rotational symmetry (C_2) but lacks reflective symmetry and, moreover, that the *cis* isomer possesses reflective symmetry (σ) but lacks rotational symmetry. Although it may be argued that the projections of Figure II-18 have limited validity in that they bear some relationship to the time-averaged cyclohexane conformations, still, they are artificial simplifications of the actual situation outlined in the previous detailed discussions. Nevertheless, the stereochemical conclusions regarding substituted cyclohexanes that may be derived from these planar projections do coincide with the conclusions based on a detailed analysis of the stereochemical relationships actually present. The use of such artificial planar projections will result in a greatly simplified but valid analysis of the biological stereospecificities possible in such structures as the inositols of Figure II-19. (Cf. Chapter VI, pp. 186–189).

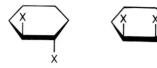

Figure II-18.   *Cis* and *trans*-1,2-disubstituted cyclohexanes—artificially simplified planar projections.

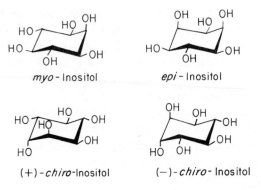

Figure II-19.   Favored conformations of some inositols.

REFERENCES

1. A. L. Lehninger, *Biochemistry*, p. 354, Worth Publishers, 1970.

2. E. E. Conn and P. K. Stumpf, *Outlines of Biochemistry*, 2nd ed., p. 143, Wiley, 1966.

3. M. Zeldin, *J. Chem. Educ.*, **43,** 17 (1966).

4. M. M. Orchin and H. H. Jaffé, *J. Chem. Educ.*, **47,** 372 (1970).

5. H. Hirschmann in *Comprehensive Biochemistry*, Vol. 12, Chapter VII, p. 258, M. Florkin and E. H. Stotz, Eds., Elsevier, 1964.

6. K. Mislow and M. Raban, *Topics in Stereochemistry*, Vol. 1, p. 7, N. L. Allinger and E. L. Eliel, Eds., Wiley, 1967.

Chapter III

MOLECULAR DISSYMMETRY, OPTICAL ACTIVITY, AND BIOLOGICAL STEREOSPECIFICITY

Thus, how can we refuse to admit that a right body has a possible left form, knowing as we do the significance of the right or left character? We might as well doubt that an irregular tetrahedron has its inverse, that a right helix has its left form, that a right hand has a possible left (L. Pasteur, 1860).[1]

ENANTIOMERIC AND DIASTEREOMERIC STEREOISOMERS

With the possible exception of vampires, all objects have mirror images. In the case of optically active molecules, however, the mirror image form is not superimposable upon the original. This is the fundamental criterion for optical activity. In analogy to the nonsuperimposable mirror image relationship of right and left hands, optically active structures may be characterized as *chiral*—that is, they possess handedness. (Chiral is derived from *chier*, the Greek word for hand.)

Isomeric chemical compounds are defined as species that have the same molecular formulas but differ in molecular structure. The possible differences in molecular structure may be divided into two fundamental classes. *Structural* or *constitutional* isomers differ in molecular bonding arrangements and are readily distinguished by differing *two-dimensional* structural formulas. Glucose and fructose are examples of biochemically important constitutional isomers and the structural formulas in Figure III-18 illustrate the contrasting bonding arrangements. *Stereoisomers*, in contrast, possess the same molecular bonding skeleton but differ in the absolute arrangement of the atoms in space. For this reason stereoisomers generally can be distinguished only by

39

appropriate *three-dimensional representations.** Stereoisomers, in turn, may be divided into two groups. Stereoisomers that are related as object and non-superimposable mirror image are termed *enantiomers*. Those stereoisomers not so related are called *diastereomers*. From these definitions it follows that a given stereoisomer may have only one enantiomeric isomer but may have several diastereomeric isomers. It also follows that two stereoisomers cannot possess both enantiomeric and diastereomeric relationships with each other. D-Glucose and L-glucose are enantiomeric stereoisomers, whereas D-glucose and D-galactose are two biologically important diastereomers (cf. Figure III-18).

Owing to the free rotations about single bonds discussed earlier, molecules may exist in a variety of geometrical forms. As defined here, however, stereoisomers differ in the *absolute* arrangement of atoms in space. The observation that an enantiomeric or a diastereomeric stereoisomer can exist in a variety of *conformations* (three-dimensional geometries) does not destroy the fundamental geometrical differences that distinguish these stereoisomeric structures. It is not possible to interconvert distinct enantiomeric or diastereomeric isomers by free rotations about the single bonds (cf. pp. 33–36). In the following descriptions it will become clear that enantiomers exist in equivalent relative conformations that differ *only* in an absolute geometrical sense. As we shall demonstrate, however, the conformations of diastereomeric isomers differ in both relative and absolute geometries.

When introduced at the elementary level, optical activity is often stated to be the result of an "asymmetric" carbon somewhere within a molecule. In describing the historical development of concepts concerning optical activity, we mentioned that van't Hoff and Le Bel first recognized that a tetrahedral carbon bonded to four chemically distinct groups would produce chiral molecular structures. Their insight explained the data of 1874: that 13 compounds of then established structure [e.g., lactic acid, $CH_3CH(OH)-CO_2H$, and aspartic acid, $HO_2CCH_2CH(NH_2)CO_2H$] occurred in optically active forms. Each of these 13 established molecular structures contained at least one asymmetric (C_{abcd}) carbon grouping. The observation that one asymmetric carbon atom within a molecule will result in enantiomeric forms and is thus a sufficient condition for optical activity† is illustrated in Figure III-1.

Occasionally, instead of emphasizing the relationship between asymmetric carbon atoms and enantiomeric molecular forms, it is stated that molecules lacking both points and planes of symmetry will exist in optically

---

* The *cis* and *trans* stereoisomeric forms of planar molecules are an exception to this general observation.

† Two or more asymmetric centers, however, may be so arranged as to result in non-optically active forms; cf. Figures II-13 and III-20.

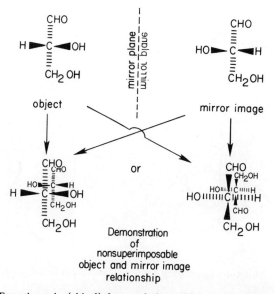

Figure III-1. Enantiomeric (chiral) forms of glyceraldehyde resulting from tne presence of one tetrahedral carbon atom bound to four chemically distinct groups.

active forms. The exact relationship between molecular symmetries and optical activity, however, is that summarized in Table II-2. In order to exist in enantiomeric forms, a molecule need not lack *all* symmetry (i.e., be asymmetric). Enantiomeric molecules lack only one particular class of symmetry—*reflective symmetry*, which we define as all improper axes of rotation.

As outlined in Table II-2, molecules lacking reflective symmetry (no S_n, $n \geq 1$) are designated as *dissymmetric*. It should be noted that *asymmetric molecules* (molecules possessing only the trivial C_1 symmetry element) *are a special class of dissymmetric molecules, but not all dissymmetric molecules will be asymmetric*. A knowledge of this more complete and exact relationship existing between molecular symmetry and optical activity permits us to readily understand the optical properties of the four molecules in Figure III-2. Molecules 1, 2, and 3 all lack asymmetric carbon atoms. Furthermore, they cannot be classified as asymmetric structures, since they all possess C_2 axes of rotation. However, they all lack S_n axes and are therefore dissymmetric. They will occur in optically active enantiomeric forms. Molecule 4, on the other hand, is not capable of existing in optically active forms. Although structure 4 lacks both points and planes of symmetry (S_1 and S_2), it possesses an S_4 improper axis of rotation. It is not, therefore, dissymmetric

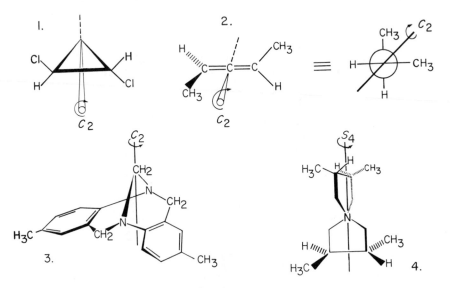

Figure III-2.   Molecules illustrating further relationships between optical activity and molecular symmetry.

and the original molecule will be superimposable upon its mirror image. Finally, Figure II-13 illustrated an instance of a dissymmetric molecule that is not optically active because of conformational racemization.

We have already defined the conformation of a molecule as its particular three-dimensional geometry or shape. Conformations may be precisely defined in terms of the bond distances, the bond angles and the dihedral angles between the adjacent groups (cf. Figure II-11). Enantiomeric stereoisomers, such as the glyceraldehyde structures in Figure III-1 have, of course, the same bond lengths and bond angles. After careful examination of molecular models, it should be apparent that enantiomers are also characterized by the same distances and angles between the various chemical groups. This observation is clearly manifest in the area of molecular crystallography. The normal techniques of X-ray crystallography, which enable the spectroscopist to *precisely determine all bond distances and angles and all intramolecular distances and angles, still will not permit differentiation between two enantiomeric possibilities.*

In contrast to enantiomers with their equivalent relative conformations, stereoisomers that are diastereomeric possess chemical groupings of distinctly different geometries. These important principles are illustrated by the 1,2-dichlorocyclopropane stereoisomers of Figure III-3. It can be seen that the enantiomeric forms resulting from the dissymmetric *trans* structure possess the same relative geometry. In both enantiomers, for example, the distances

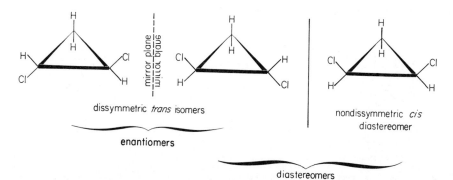

Figure III-3.   1,2-Dichlorocyclopropane stereoisomers.

between the two *trans* chlorine atoms are exactly the same, and the relative orientations of the chlorine atoms with respect to all other atoms of the molecule are exactly the same. Only in an absolute sense, with reference to an arbitrary standard—right-handed or left-handed, clockwise or counter-clockwise—do the geometries of these two stereoisomers differ. The non-dissymmetric *cis*-1,2-dichlorocyclopropane stereoisomer is diastereomeric to both of the *trans* isomers. It is readily apparent that the distances between the two chlorine atoms and the relative orientation of these two atoms differ markedly in the *cis* and *trans* diastereomeric structures. The differences in the relative arrangement of the similar chemical groups in diastereomeric structures will be reflected in differences in their chemical and physical pro-perties. For example, in the present illustration, the dipoles due to the polar carbon–chlorine bonds will be additive in the *cis* isomer but will tend to cancel each other in the *trans* isomeric forms. Thus the net molecular dipole of *cis*-1,2-dichlorocyclopropane will be significantly greater than that of *trans*-1,2-dichlorocyclopropane.

It is particularly apparent that diastereomeric structures have different relative geometries in the conformationally rigid cyclopropane system. This remains true, although less readily illustrated, in diastereomeric structures that are able to adopt different conformations. The somewhat more flexible cyclohexane system may be used to study the situation that develops as greater conformational freedom is allowed. Inspection of the disubstituted cyclohexane isomers in Figure II-15 reveals that even the *average* X–X orientations of the *cis* isomer will differ from the *average* X–X orientations in the diastereomeric *trans* structure.

The Newman projections in Figure III-4 of various conformational isomers derived by free rotations about the C-2–C-3 bonds of the enantio-

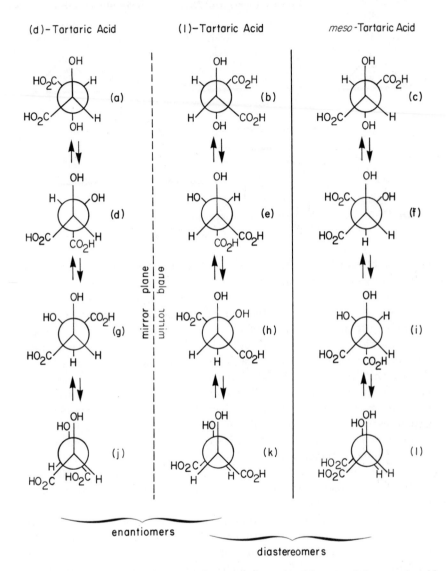

Figure III-4. Newman projections of some conformational isomers of the tartaric acid stereoisomers.

meric dextrorotatory and levorotatory tartaric acids and of the diastereomeric *meso*-tartaric acid illustrate the same principle. There are no conformations, for example, in which the two secondary alcohol groups of the *meso*-acid and the corresponding alcohol groups of the *d,l*-acids are found to be in the same relative chemical and physical environment. For example, when the two hydroxyl groups of the optically active tartaric acids are exactly opposed to each other, they experience an unbalanced environment, since both carboxyl functions are then located on one side of the plane determined by the two hydroxyl groups and the two interior carbon atoms (C-2 and C-3). In the *meso* structure, on the other hand, when the two hydroxyl groups have an exactly opposed relative orientation, then one carboxyl group and one hydrogen atom is found on each side of this plane [projections (a) and (b) versus (c)]. When the two hydroxyl groups of the *d* and *l* isomers eclipse each other, one carboxyl group and one hydrogen atom are now found on each side of the $HO-C_2-C_3-OH$ plane; but when the two hydroxyl groups of the *meso* form are in this eclipsing relationship, then the carboxyls and hydrogens are also eclipsing each other and therefore like groups fall on one side of this plane [projections (j) and (k) versus (l)]. When the two hydroxyl groups of the *d,l* acids are oriented so that the dihedral angle between them is an acute angle, then the groups opposing the hydroxyl functions are found to be either the two carboxyl functions [projections (d) and (e)] or the two hydrogen atoms [projections (g) and (h)]. In contrast, when the two hydroxyl groups of the *meso* form have this relative orientation, then [projections (f) and (i)] the opposing groups to the hydroxyls are carboxyl *and* hydrogen. These observations serve to illustrate that even when free rotations about carbon–carbon single bonds are considered, diastereomeric structures will possess different relative geometries or conformations. In the case of the tartaric acid conformations, this results in measurably different ionization constants for the diastereomeric acids (For *dl*-tartaric acid $pK_1 = 2.9$ and $pK_2 = 4.2$; for *meso*-tartaric acid $pK_1 = 3.1$ and $pK_2 = 4.8$).[2]

NMR DIFFERENTIATION OF DIASTEREOTOPIC PROTONS

The technique of nuclear magnetic resonance (nmr) spectroscopy is a powerful and useful means of detecting chemically distinct groups. In the presence of an externally applied magnetic field, many nuclei (including ^1_1H, $^{13}_6C$, $^{19}_9F$, and $^{31}_{15}P$)* undergo a radiofrequency-induced transition to a

---

* The numerical superscript and subscript preceding the symbol for a chemical element indicate particular isotopic forms of the elements. The superscript represents the atomic mass; the subscript the atomic number. Thus ^2_1H is a deuterium atom and ^3_1H is a tritium atom. For convenience, however, we frequently use the symbols D (for deuterium) and T (for tritium) to represent ^2_1H and ^3_1H, respectively.

higher energy level. The energy absorbed during this transition is detected by a separate radiofrequency detector circuit. Although in a given magnetic field each type of atomic nucleus should absorb energy of equivalent frequency, in fact, each nucleus is slightly shielded from the applied magnetic field by its surrounding electron cloud. Since the density of this electron cloud will vary with the chemical environment of the nucleus, the resonance frequencies of different atomic nuclei within a molecule will vary, even if they belong to the same isotopic class. For example, the nmr frequencies for the protons of a given molecule will display different chemical shifts which reflect the differing chemical environments of these protons.*

The nicotinamide adenine dinucleotides (NAD and NADP) are important participants in many biological oxidation-reduction reactions. Dehydrogenase enzymes catalyze the reversible transfer of a hydride ion from various substrates to the C-4' position of the nicotinamide ring of these cofactors to produce the NADH and NADPH structures in Figure III-5. These catalyzed hydride transfers manifest biological stereospecificity, since the enzymes differentiate between the paired geminal hydrogen atoms at C-4' of the dihydronicotinamide rings (cf. Chapter VI, pp. 240–244).

Although the paired hydrogens at C-4' of the nicotinamide are *chemically like*, a consideration of the total three-dimensional structures of NADH and NADPH projected in Figure III-5 reveals that these hydrogens *cannot be interchanged by any symmetry operation.* The presence of several asymmetric centers (at C-1, C-2, C-3, and C-4 of the D-ribofuranosyl units) with particular configurations dictates that these paired C-4' hydrogens always be situated in geometrically and chemically different environments. This will be true even when conformational changes (free single bond rotations, umbrella inversions at the tertiary nitrogens, and interconversions of the dihydronicotinamide ring geometries) of these molecules are considered. The chemical and geometrical relationships between these C-4' paired hydrogen atoms are therefore analogous to those between the diastereomeric tartaric acid structures described previously. For these reasons, the paired hydrogens at C-4' of the dihydronicotinamide rings of NADH and NADPH may be designated as *diastereotopic paired groups* (Table II-3).

Since these diastereotopic hydrogens are situated in chemically differing environments, the nmr spectra of NADH should reveal distinct resonance frequencies for each of the C-4' hydrogens. The initial nmr investigations using 60-MHz [60 × 10⁶ Hz radiofrequency excitation with an external magnetic field of about 14,100 G (gauss)] or 100-MHz (100 × 10⁶ Hz with

---

* Excellent nonmathematical introductions to nmr are provided by L. M. Jackman (*Applications of Nuclear Magnetic Resonance Spectroscopy in Organic Chemistry*, Pergamon, 1959) and J. D. Roberts (*Nuclear Magnetic Resonance Applications to Organic Chemistry*, McGraw-Hill, 1959).

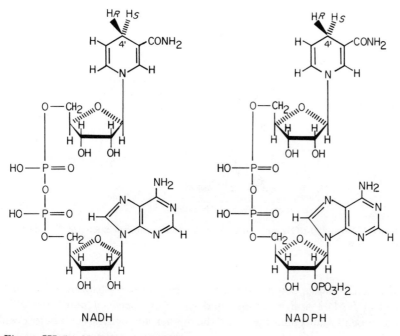

Figure III-5. NADH and NADPH.

a 23,500 G field) instruments revealed a single resonance absorption for both these hydrogen atoms. However, the difference in nmr absorption frequencies for chemically distinct protons (chemical shift) is directly proportional to the strength of the externally applied magnetic field, and the absorption frequencies of these diastereotopic paired C-4′ hydrogen nuclei have now been successfully resolved[3,4] using the recently developed 220-MHz instrument (external field applied is about 52,000 G). The resolved spectra of these nuclei appear in Figure III-6.

That a 220-MHz nmr spectrometer is required to resolve the separate chemical shifts of the paired C-4′ hydrogens of the dihydronicotinamide ring is an indication of the slight chemical difference existing between these two atoms. As illustrated in Figure III-6, Patel found that the difference in chemical shifts for these two hydrogens is temperature dependent. At 16° the resonance frequencies of these paired hydrogens are separated by about 0.065 ppm, but at 59° the chemical shift difference is too slight to permit resolution of the separate absorption frequencies. Patel ascribed this circumstance to a decrease in the amount of specific positioning of one of these nonequivalent hydrogen atoms next to the adenine ring at higher tempera-

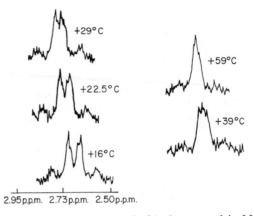

2.95 p.p.m.   2.73 p.p.m.   2.50 p.p.m.

Figure III-6.   220-MHz nmr spectra of the C-4′ hydrogen nuclei of NADH in D_2O, pD = 9.0.[4]

tures. This change could result from a greater concentration of the "open" form of NADH and/or more rapid equilibria between two "folded" and two "open" molecular forms at the higher temperatures. These possible equilibria are outlined in Figure III-7. Other nmr evidence also suggests that the pyridine nucleotides can exist in folded conformations with the adenine and pyridine rings in associated parallel planes.[5] The explanation offered for the observed temperature effect is also consistent with the fact that the reduced nicotinamide mononucleotide structure, which lacks the adenosine moiety (NMNH), displays a *single resonance* for the C-4′ paired geminal hydrogens of the dihydronicotinamide ring even in a 220-MHz spectrometer (Figure III-8).

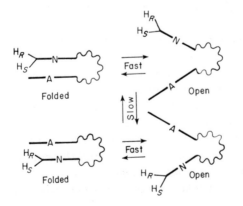

Figure III-7.   Potential equilibria between folded and open conformations of NADH.[4]

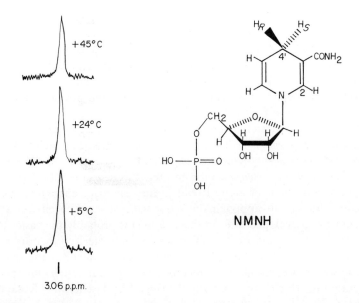

Figure III-8. Temperature dependence of the 220-MHz nmr spectrum of NMNH C-4′ hydrogens in D_2O, pD = 9.2.[4]

These nmr investigations of NADH illustrate several important points. They show conclusively that the C-4′ paired hydrogens of the dihydronicotinamide ring of NADH and NADPH are distinct and can be differentiated by sophisticated physicochemical measurements. Thus these diastereotopic paired atoms within a given molecule are seen to be analogous to separate diastereomeric molecules. They experience chemically and geometrically distinct environments. In the case of the C-4′ hydrogens of NADH, the chemical differences are so slight that even with a 220-MHz nmr spectrometer the two absorptions cannot be resolved *unless* the chemical differences are *enhanced* by a stereospecific interaction with the associated adenine ring. It would be a serious error, however, to conclude that the *enzymatic differentiation* between the paired geminal hydrogens at C-4′ of the dihydronicotinamide rings in NADH and NADPH is totally dependent on the same stereospecific interaction with the adenine rings. The diastereomeric C-4′ hydrogens of the dihydronicotinamide rings are chemically and geometrically nonequivalent in NMNH *and* in the "open" conformations of NADH, as well as in the "folded" NADH conformations. It has *only* been demonstrated that the chemical differences between the paired C-4′ dihydronicotinamide hydrogens of NMNH and "open" NADH are not sufficient to be detected by 220-MHz nmr spectroscopy. This book presents many examples of biological stereospecificity between enantiotopic paired groups that are situated in *equivalent relative chemical environments* and *differ only* in their *absolute geometries and absolute chemical environments*. (See following discussion.) There is no inherent reason, therefore, why biological differentiation between the C-4′ paired hydrogens of the dihydronicotinamide ring of NMNH or the "open" conformation of NADH could not be observed with a suitable enzymatic system.

In contrast to diastereomeric molecules, *enantiomers possess the same relative three-dimensional geometry and*, except in a dissymmetric environment, *are characterized by equivalent physicochemical properties. Enantiomers do, however, possess opposite configurations. Configuration* is the term used to describe the absolute order of arrangement of atoms about a dissymmetric center or region of a molecule. Practice in applying the sequence system of specifying absolute configuration (developed later in this chapter) is probably the easiest way to demonstrate that enantiomeric molecules are characterized by opposite configurations. Before describing the sequence method of specifying configurations, however, we discuss the older but less general system for specifying configurations which was originally developed by Fischer and Rosanoff.

FISCHER-ROSANOFF ABSOLUTE CONFIGURATIONAL DESIGNATIONS: THE D AND L SYSTEM

Fischer's investigations of biological compounds pointed up the need for a standardized system of configurational nomenclature. In order to correlate the structures of the aldohexoses, for example, it became necessary to describe how the absolute configurations of the glucose and galactose carbon atoms differed. One solution to the problem of configurational nomenclature would be to draw clear three-dimensional perspectives of all dissymmetric molecules. This method is always valid, and the perspective drawings of (+)-glucose and (+)-galactose in Figure III-9 can be used either to compare the configurations between corresponding asymmetric centers in these molecules or to correlate these configurations with the asymmetric centers of any other molecule. It is obvious, however, that this method of specifying configurations is not suited for molecular descriptions that must appear in normal type or script. In order to communicate the chirality of a set of molecules, this method of specifying configurations requires a prohibitive number of perspective drawings. It is thus used sparingly in scientific papers and books.

Since enantiomers differ in the direction in which they rotate the plane of polarized light, it would seem that a second obvious method of specifying and correlating configurations would be simply to indicate the direction a particular dissymmetric molecule rotated polarized light. Unfortunately

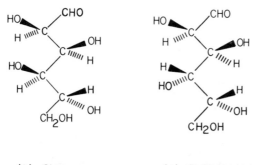

(+)- Glucose   (+)-Galactose

Figure III-9.   Perspective drawings of the natural glucose and galactose stereoisomers.

however, although the dissymmetry of a molecule is directly responsible for the observation of optical activity, no regular correlation can be specified between the direction of rotation of polarized light and the absolute configuration of the dissymmetric center. For example, Fischer was aware that the esterification of dextrorotatory lactic acid [(+)-lactic acid] resulted in a levorotatory methyl ester [methyl (−)-lactate]. Since esterification of the carboxyl group of lactic acid does not reverse the configuration at the asymmetric carbon atom, this example serves to illustrate that the sign of rotation of a molecule is not a valid characterization of the absolute order of arrangement of groups about a dissymmetric center—even within a closely related series of compounds.

Since the first possible method of designating configurations discussed here was impractical and the second possible method was invalid, Fischer chose a third procedure—relating the configurations of dissymmetric compounds to a defined standard configuration.

As the result of extensive chemical reactions carried out with the various hexose sugars, for example, Fischer was able to correlate the relative configurations of all the asymmetric centers within this group of compounds. By *arbitrarily* choosing an absolute configuration for one standard reference compound, then the absolute configurations of *all* the hexoses were established—relative to this original arbitrary standard. Fischer selected the configuration in Figure III-10 to represent the absolute configuration of the saccharic acid (glucaric acid) derived by oxidation of normal (+)-glucose. Using this as a standard reference configuration, Fischer's correlation system of configurational designations permitted classification of the carbohydrates into only two stereochemical groups. One stereochemical series, represented by normal glucose and fructose, Fischer designated by *d*. The alternative

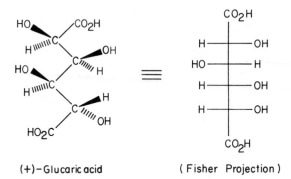

$$=$$

(+)−Glucaric acid   (Fisher Projection)

Figure III-10. Absolute configuration of dextrorotatory glucaric acid as *arbitrarily* assigned by Fischer.

stereochemical series was designated with an *l*.* Despite the initial success of this stereochemical classification scheme, Fischer's system of designating configurations contained some fundamental errors and it did not survive in its original form. The original choice of (+)-glucaric acid as the configurational reference standard resulted in attempts to directly relate the *total* configurations of the hexoses. Direct correlations of the hexoses possessing multiple centers of dissymmetry through the saccharic acids also possessing multiple centers of dissymmetry led to several serious inconsistencies. For example, natural xylose, assigned by Fischer to the *l* series, could be changed to *d*-lyxose. Although some investigators pessimistically concluded that a perfectly consistent configurational classification of the sugars could not be developed, in 1906 Rosanoff, using the stereochemical configurations proposed by Fischer, outlined a modified system of configurational nomenclature for the carbohydrates that eliminated the inconsistencies in Fischer's original *d* and *l* classifications.[6] Rosanoff's modification of Fischer's *d* and *l* configurational system is still used today.

Rosanoff examined the three-dimensional representations of the carbohydrates and recognized that a systematic designation of configurations based on the absolute configuration of a *single specified dissymmetric center* in each molecule was possible. Although the glyceraldehydes had not yet been isolated and correlated with the higher sugars, Rosanoff proposed that the absolute configuration of the single dissymmetric center of the glyceroses (glyceraldehydes) be established as the standard configurational reference, replacing Fischer's glucaric acid standard. When the configuration at the

* Cf. E. Fischer, "Nobel Lecture, December 12, 1902," in *Nobel Lectures in Chemistry*, p. 32, Elsevier, 1966.

penultimate carbon (next to the highest numbered carbon atom when the saccharide chain is numbered from the aldehydic carbon) of each carbohydrate was compared with the glyceraldehyde standard configuration, Rosanoff showed that all the carbohydrates could be divided into two consistent stereochemical classifications. One group of carbohydrates could be designated as possessing the configuration of the dextrorotatory glyceraldehyde reference standard at the penultimate carbon atom. This classification corresponded generally to the compounds Fischer had classified within the *d* configurational grouping. The other group of carbohydrates could be designated as possessing the configuration of the enantiomeric levorotatory glyceraldehyde isomer at the penultimate carbon atom and corresponded generally to Fischer's *l* configurational grouping. The schematic representation in Figure III-11, reproduced from Rosanoff's 1906 paper, summarizes his proposed configurational designations.

In Figure III-11 the sugars are written in abbreviated "Fischer projections." * In this shorthand notation the bonds to the hydroxyl groups are represented by the short bars perpendicular to the axis of the carbon chain of the sugar in the Fischer projection and the bonds to the hydrogen atoms are omitted (Figure III-12). Such schematic representations of Fischer projections were also first proposed by Rosanoff in this paper.

The proposed configurational scheme of Figure III-11 is divided into two semicircular groupings. All the compounds placed by Rosanoff on the right side of the vertical dividing line possess the same absolute configuration at the penultimate carbon atoms as the reference standard "glycerose" with the hydroxyl projected to the right. This "glycerose" was later correlated with natural dextrorotatory glyceraldehyde. Those compounds placed within the left semicircle all possess the enantiomeric configuration (hydroxyl group projected to the left) at the penultimate carbon atoms. Those carbohydrates originally classified by Fischer as belonging to the *d* series are designated in Rosanoff's diagram by D, and the carbohydrates originally classified by Fischer as *l* are designated by L. It is readily apparent how (and why) Rosanoff's modification resulted in the reclassification of several of the carbohydrate structures. In the text of his paper, Rosanoff proposed that all the compounds falling in the right-hand portion of the diagram be designated as possessing the δ configuration and those in the left-hand portion of the diagram be designated as possessing a λ configuration. Although the Rosanoff *classification* of Figure III-11 became standard, the δ and λ *designations* were never widely accepted. Instead, we now use D to designate the absolute configurations of *all* the compounds on the right side of the diagram

---

* Cf. pp. 57–61 for a description of the Fischer convention for representing carbohydrate structures in two-dimensional projections.

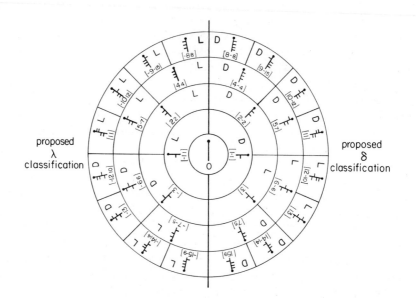

proposed
λ
classification

proposed
δ
classification

	Acids		Sugars		Alcohols
1.	Oxalic	0.	Glycolose; dioxyacetone	0.	Ethyleneglycol
1,–1.	Tartronic	1.	Glycerose; erythrulose	1,–1.	Glycerol
2,–2.	*Meso*-tartaric	2.	Erythrose; araboketose	2,–2.	*Meso*-erythrite
3.	Tartaric	3.	Threose; xyloketose	3.	Erythrite
4,–4.	Ribotrioxyglutaria	4.	Ribose	4,–4.	Adonite
5,7.	Trioxyglutaric	5.	Arabinose; fructose	5,7.	Arabite
6,–6.	Xylotrioxyglutaric	6.	Xylose; sorbinose	6,–6.	Xylite
7,5.	=5,7	7.	Lyxose; tagatose	7,5.	=5,7
8,–8.	Allomucie	8.	Allose	8,–8.	—
9,15.	Talomucic	9.	?	9,15.	Talite
10,–12.	+Saccharic	10.	Glucose	10,–12.	−Sorbite
11.	Mannosaccharic	11.	Mannose	11.	Mannite
12,–10.	−Saccharic	12.	Gulose	12,–10.	+Sorbite
13.	Idosaccharic	13.	Idose	13.	Idite
14,–14.	Mucie	14.	Galactose	14,–14.	Dulcite
15,9.	=9,15	15.	Talose	15,9.	=9,15

Figure III-11. Rosanoff's proposed configurational classification of the carbohydrates.[6]

54

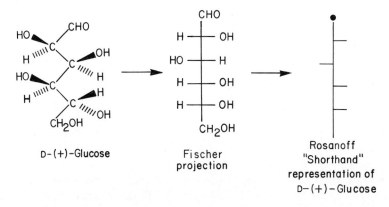

D-(+)-Glucose   Fischer projection   Rosanoff "Shorthand" representation of D-(+)-Glucose

Figure III-12.   Rosanoff's representation of a Fischer projection of D-(+)-glucose.

and L to designate the enantiomeric carbohydrate configurations on the left side of the diagram.*

As evolved into its current usage, then, the system of designating configurations developed by Fischer and Rosanoff is based on the standard reference configuration of dextrorotatory glyceraldehyde [or (+)-glyceraldehyde] represented in Figure III-13. Consistent with the previously given nomenclature this standard reference configuration is designated by the small capital D, and the opposite (mirror image) configuration by the small capital L. The reference compound selected by Rosanoff—with the absolute configuration arbitrarily chosen by Fischer—is thus D-(+)-glyceraldehyde.† This compound can also be termed D-(*d*)-glyceraldehyde. In this case the D

---

* The situation must certainly be described as potentially confusing. The δ and λ designations of Rosanoff are no longer used. The compounds corresponding to the proposed δ classification—the right side of Fig. III-11—are now designated with a D, the symbol used by Rosanoff in his schematic diagram to indicate those compounds assigned as *d* by Fischer. Similarly, compounds of the proposed λ configuration—the left side of Fig. III-11—are now designated with L, the symbol used by Rosanoff to indicate compounds classed as *l* by Fischer. Today the *d* and *l* designation originally proposed as configurational classifications by Fischer are used interchangeably with (+) and (−) to indicate dextrorotatory or levorotatory optical activity, respectively.

† Although the original choice of the configuration that would be assigned to (+)-glyceraldehyde was arbitrary, the selection was an inspired one. Sixty years after the original designation by Fischer, J. M. Bijvoet et al.,[7] using a special technique of X-ray spectroscopy, demonstrated that the compounds belonging to the D series do in fact possess the absolute order of arrangement of substituents depicted in the D-(+)-glyceraldehyde projection of Figure III-13. Thus we now know that all the configurations that had been established relative to that of the original arbitrary standard are correct absolute configurations.

Figure III-13.   Dextrorotatory glyceraldehyde ≡ D-(+)-glyceraldehyde.

indicates the configurational designation and *d* or (+) indicates the *incidental* observation that the compound is dextrorotatory.

To designate configurations according to the Fischer-Rosanoff system, the absolute orientation of a dissymmetric center is correlated with the D-(+)-glyceraldehyde standard configuration. If the absolute three-dimensional arrangement of the dissymmetric center is known, the assignment of the D or L absolute configuration can be made from inspection of models. In general, however, the correlation must be made by chemically converting the compound with a dissymmetric center of unknown configuration into a molecule that can be (or has previously been) directly compared with D-(+)-glyceraldehyde. For example, when the aldehydic carbonyl of D-(+)-glyceraldehyde is oxidized to a carboxyl function, (−)-glyceric acid is produced. The asymmetric center of the levorotatory glyceric acid therefore has the same absolute configuration as does D-glyceraldehyde and can be designated as D-(−)-glyceric acid. Oxidation of the carbonyl group and reduction of the hydroxymethyl group of glyceraldehyde produces lactic acid. Experimental observations have established that levorotatory lactic acid can be correlated with the standard D-(+)-glyceraldehyde configuration and is therefore D-(−)-lactic acid. The observation that (+)-lactic acid yields a (−)-lactate methyl ester should therefore be written as

L-(+)-lactic acid → methyl L-(−)-lactate

to indicate that, although the direction of rotation of polarized light changes, the absolute configuration of the lactate molecule remains the same. The configuration of the common form of alanine, (+)-alanine, can be correlated with the glyceraldehyde standard through a correlation with lactic acid. In comparing these configurations, the amino group of alanine is considered to correspond to the hydroxyl function of lactic acid [and of the original reference D-(+)-glyceraldehyde]. Since it was possible to correlate the configuration of L-(+)-lactic acid with (+)-alanine, natural dextrorotatory alanine is L-(+)-alanine.[8]

Through a series of synthetic and degradative reaction sequences in which any changes in the configuration of the asymmetric centers were care-

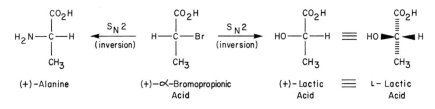

Figure III-14. Correlation of the L-(+)-lactic acid configuration with (+)-alanine.[8]

fully cataloged,* it has been possible to correlate the configurations of many enantiomeric molecules with the absolute D-(+)-glyceraldehyde configurational standard. It is now realized that the normal form of the α-amino acids possesses the same absolute configuration at the α-carbon atom as the L-(−)-glyceraldehyde standard, and therefore all belong to the L-series.

Many molecules of biochemical interest possess more than one asymmetric carbon atom. When the absolute configuration of a single asymmetric center in such a molecule is established, knowledge of the *relative* intramolecular configurations automatically results in establishing the absolute configuration of all centers relative to the D or L reference standards. The Fischer-Rosanoff system of designating absolute configurations can thus be applied to molecules with multiple dissymmetric centers by specifically designating one single center to be directly correlated with the reference configuration. As described previously, Rosanoff illustrated how correlation of the configuration at the penultimate carbon atom of the carbohydrates could be used to assign these molecules to consistent stereochemical classes. In the case of another important group of biological compounds, the amino acids, the configuration at the α-carbon center is specifically correlated with the D-(+)-glyceraldehyde configuration to establish the appropriate D or L configurational classification. The unnatural amino acid, D-(+)-serine, which has been correlated with the original D-(+)-glyceraldehyde configuration, is now commonly used as a more convenient secondary reference standard with compounds related to the α-amino acids (cf. p. 67 following).

For convenience in writing and printing, Fischer established the convention for projecting a three-dimensional molecule onto a page that is illustrated in Figure III-15. In a Fischer projection, the molecule is first oriented with the carbon chain vertical and to the rear, and the edge of the tetrahedron joining the hydrogen and hydroxyl function (or their equivalent groups) horizontal and to the front. The groups are then projected onto the

* The conversion of (+)-α-bromopropionic acid to L-(+)-lactic acid and to (+)-alanine in Figure III-14, for example, involved two separate inversions of the asymmetric center, for a net retention of configuration rather than a simple overall retention.

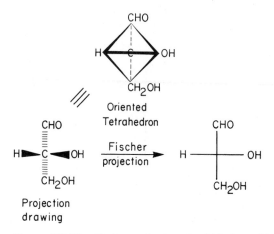

Figure III-15.   Fischer projection of D-(+)-glyceraldehyde.

plane of the page with the order of the groups as they appear in the oriented tetrahedral model indicated by the projection line drawing. The asymmetric carbon is indicated in the projection by the intersection of lines representing the projected vertical and horizontal edges of the original tetrahedral model. If the carbon chain contains two or more asymmetric centers, the projection is made in the same manner from a model oriented with the formyl (CHO) or equivalent group at the top and the hydroxymethyl (CH₂OH) or equivalent group at the bottom, the carbon chain vertical and to the rear, and the edges of the tetrahedron joining the hydrogen and hydroxyl groups horizontal and to the front. Because the Fischer projection is a defined convention for representing a *three-dimensional molecular structure in two dimensions*, only movements of the projection line drawing that keep all points of the drawing confined to the plane of the page are permitted. Thus we may translate the projection, or rotate it about an axis perpendicular to the page. Rotation of Fischer projections about axes lying in the plane of the page, however, result in a change in the orientation of the original tetrahedral model such that it is no longer properly positioned for the specified Fischer projection. Obviously, *we can rotate the original three-dimensional molecular model* in order to gain a different perspective, *but the rotation of the Fischer projection* in a manner that lifts it out of the plane of the paper *leads to a line drawing that is no longer a valid Fischer projection*. These aspects of Fischer projections are presented in Figure III-16.

One valid and useful manipulation of Fischer projections does not fit into the previously described categories. It is possible to rotate a tetrahedral carbon grouping about an axis defined by the bond joining the central carbon

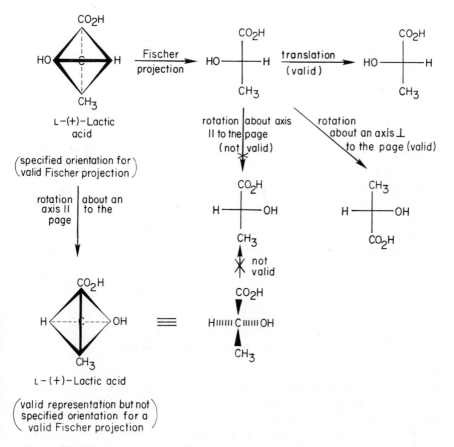

Figure III-16. Valid and nonvalid manipulations of a Fischer projection.

to any atom attached at one of the vertices. Rotation about one of these four possible axes causes a regular clockwise or counterclockwise progression of the groups attached at the remaining three vertices, without destroying the specified molecular orientation for a valid Fischer projection. Thus it is permissible to rotate any *three* attached groups of a Fischer projection in a regular and stepwise fashion while keeping the position of the central carbon and the remaining attached group fixed. This operation is often helpful in establishing the correct correlation between a Fischer projection and the *R* and *S*, or sequence method of specifying absolute configuration. Such manipulations are displayed in Figure III-17.

The Fischer projection formulas in Figure III-18 illustrate the total absolute configurations established for the eight aldohexoses now known to

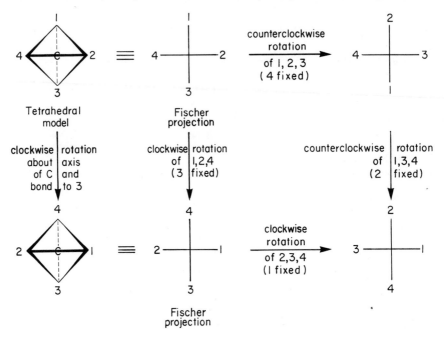

Figure III-17. Additional valid manipulations of Fischer projections.

belong to the D classification. The naturally occurring tetroses and pentoses are biosynthetically related to the natural D-hexoses; and by applying the convention that the configuration at the penultimate carbon of the carbohydrate determines the stereochemical series, we realize that these sugars also are members of the D series. The common ketohexose, D-fructose, and some D-tetroses and D-pentoses are also indicated in Fischer projection in Figure III-18. In each case the enantiomeric mirror image isomer, because it will possess the opposite or L-configuration at the penultimate carbon, will be the corresponding L-carbohydrate.

In this manner the Fischer-Rosanoff classifications may be used to successfully characterize the absolute configurations within the carbohydrate series. It should be apparent from the structures of Figure III-18, however, that the D or L configurational classification of the carbohydrates does *not* explicitly described the configuration of each dissymmetric center in these molecules. It is necessary to know (or look up) the configurations relative to the explicitly classified penultimate carbon configuration in order to totally described a D-glucose or D-ribose molecule. This is a serious deficiency of the D and L configurational classification when applied to structures containing multiple dissymmetric centers.

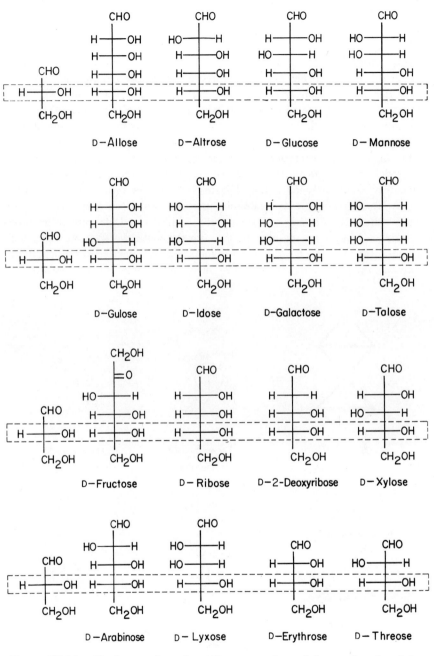

Figure III-18. Absolute configurations of some members of the D-series of carbohydrates.

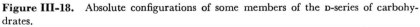

The two aldotetroses indicated in Figure III-18 may be used to illustrate some important aspects of molecules that contain two similar asymmetric carbon atoms. In Figure III-19 the original three-dimensional model from which the Fischer projections of D-threose and D-erythrose of Figure III-18 were derived are converted into the corresponding Newman projections. As can be readily seen in the Newman projection of an appropriate conformational form, the erythrose stereoisomer allows corresponding groups on the two adjacent asymmetric carbon atoms to be simultaneously eclipsing each other. The diastereomeric threose stereoisomer, in contrast, cannot be rotated into a conformation in which the corresponding groups are simultaneously eclipsing each other. Diastereomeric stereoisomers having configurational arrangements analogous to those of threose and erythrose are often indicated by a *threo* or *erythro* prefix, as illustrated for some 3-bromo-2-butanol stereoisomers. Neither the *threo* nor the *erythro* designation will specify a single stereoisomer. Examination of models or perspective drawings should convince the reader that the L enantiomers of threose and erythrose will also

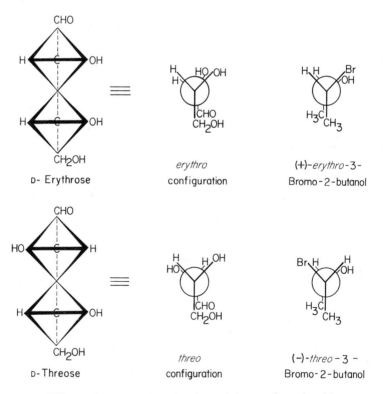

Figure III-19.   Representation of *erythro* and *threo* configurational isomers.

exist in *threo* and *erythro* configurations, respectively. The particular enantiomeric forms of the *threo* and *erythro*-3-bromo-2-butanols pictured in Figure III-19 are specified by the signs of rotation.

If the D-threose and D-erythrose molecules are oxidized with nitric acid, the formyl groups and the hydroxymethyl groups are both converted to carboxylic acid functions, to produce diastereomeric (not enantiomeric) tartaric acid molecules. As illustrated in Figure III-20, the product of D-threose is the optically active, levorotatory, tartaric acid. We will designate this tartaric acid molecule D-(−)-tartaric acid (but see following discussion). Oxidation of L-threose would produce the enantiomeric L-(+)-tartaric acid. Oxidation of D or of L-erythrose, however, produces *superimposable* mirror image forms. These projections, therefore, represent a single nondissymmetric tartaric acid stereoisomer in which there are two comparable asymmetric centers of opposite (mirror image) configuration. This form of tartaric acid possesses reflective symmetry ($\sigma = S_1$) and is the optically inactive *meso*-tartaric acid. The relationships, summarized in Figure III-20, indicate the correspondence that exists between the *erythro* and *meso* configurations and between the *threo* and *d,l* configurations.

Tartaric acid is the classical example of a molecule with two similar asymmetric carbon atoms. One asymmetric center is sufficient condition for optical activity, and the number of stereoisomers generally increases exponentially with the number of asymmetric centers. These observations lead to the

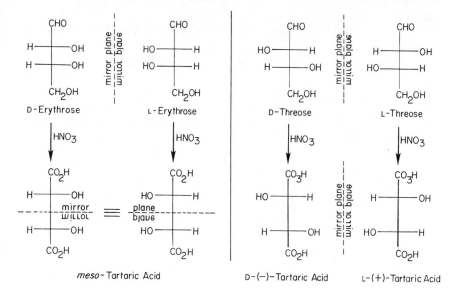

Figure III-20. Conversion of erythrose and threose into the tartaric acid stereoisomers.

general rule that the number of possible stereoisomers $= 2^n$, where n is the number of asymmetric carbon atoms in the molecular structure. This rule, however, cannot be applied blindly. We must carefully visualize the three-dimensional structures and eliminate as separate stereoisomers the superimposable mirror image forms. Thus, although the aldohexoses exist in 16 distinct stereoisomeric forms as predicted (2^n, $n = 4$) and the aldotetroses exist in four distinct stereoisomeric forms (2^n, $n = 2$), the tartaric acids, with two asymmetric carbons, exist in only three distinct stereoisomeric forms (cf. Figures III-18 and III-20). As illustrated, the two potential tartaric acids derived from the erythrose enantiomers represent only a single, optically inactive, diastereomeric form of tartaric acid.

The tartaric acid structures may also be used to illustrate a rather important limitation of the Fischer-Rosanoff method of specifying absolute configurations. We indicated that the levorotatory tartaric acid produced by oxidation of D-threose would be specified as D-(−)-tartaric acid and the enantiomer produced by oxidation of L-threose would be L-(+)-tartaric acid. This correlation appears straightforward and correct. However, the tartaric acids may also be correlated with D-(+)-glyceraldehyde through D-(−)-lactic acid and D-(+)-malic acid as shown in Figure III-21. Again the correlation is straightforward—but now we would conclude that it is the dextrorotatory tartaric acid which belongs to the D series.

Turning to correlations based on synthesis, D-(+)-glyceraldehyde may be converted into *meso* and levorotatory tartaric acid through the sequence outlined in Figure III-22.

In producing the cyanohydrin of D-(+)-glyceraldehyde, a new asymmetric center is produced. This leads, as indicated, to two diastereomeric cyanohydrin isomers. One of these, when carried through the synthetic sequence, leads to the optically inactive *meso*-tartaric acid. Since we started with D-(+)-glyceraldehyde, we can be certain that at least *one* asymmetric center in *both* the product tartaric acids corresponds to the D configuration. *Meso*-tartaric acid has a mirror plane passing between the two asymmetric

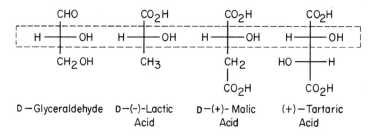

Figure III-21.   Correlation of dextrorotatory tartaric acid with D-(+)-glyceraldehyde.

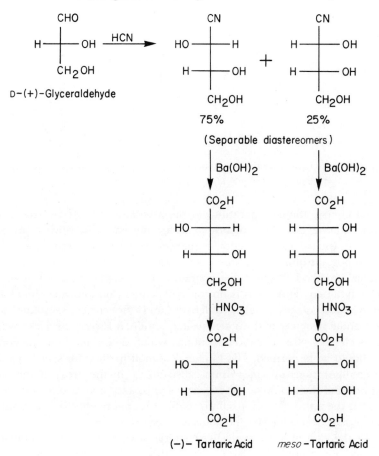

Figure III-22. Synthesis of *meso* and levorotatory tartaric acids from D-(+)-glyceralde-hyde.[9]

carbons, and the two asymmetric carbons are related as mirror images. Since one center is D, the other must possess the L configuration. The second tartaric acid product, which results from the diastereomeric cyanohydrin, must possess the opposite configuration at the newly created center, and must therefore possess D absolute configurations at *both* asymmetric centers. Again, we conclude that the levorotatory tartaric acid belongs to the D series.

However, at least in theory, we could have carried out the synthesis of *meso* and levorotatory tartaric acid from L-(+)-aspartic acid as outlined in Figure III-23. In this case we would conclude that levorotatory tartaric acid belongs to the L configurational series.

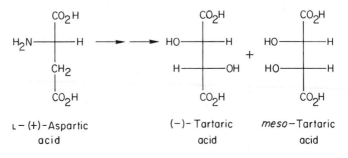

Figure III-23. Synthesis of *meso* and levorotatory tartaric acids from L-aspartic acid.

Of course, the configurations of the asymmetric centers of the dextro-rotatory and levorotatory tartaric acids are not changing during any of the foregoing discussions or reaction sequences. What these discussions indicate is an important limitation of the Fischer-Rosanoff (or any absolute reference standard) method of specifying configurations. Application of this method of specifying configurations is not only dependent on knowing the absolute configuration of D-(+)-glyceraldehyde and the correct absolute three-dimensional structure of the molecule in question, it also requires a knowledge of a specific set of arbitrary conventions which determine how the correlation should be performed. The Fischer-Rosanoff method for specifying absolute configurations operates quite satisfactorily in the areas of the carbohydrates and the amino acids. In view of this, and in view of its established position, the D,L system of absolute configurational nomenclature is unlikely to be replaced in these specific fields for some time.

In the case of a molecule like tartaric acid, however, the conventions for specifying absolute configuration by this method become arbitrary and therefore subject to confusion and controversy. If we consider tartaric acid to fall within the class of carbohydrates and their derivatives, then we will correlate the configuration at the penultimate carbon atom of the tartaric acids with that of D-(+)-glyceraldehyde, and conclude (Figures III-20 and III-22) that levorotatory tartaric acid belongs to the D series. If, on the other hand, we consider tartaric acid to fall within the class of amino acids and their derivatives, then we will correlate the configuration of the α-carbon atom of the tartaric acids with that of D-(+)-glyceraldehyde, and we will conclude (Figures III-21 and III-23) that dextrorotatory tartaric acid belongs to the D series.

In the case of molecules such as the tartaric acids, the potential confusion can be at least partially avoided by modifying the D or L configurational notation. The symbol D_g is used to denote a compound that falls within the D configurational class—when correlated with D-(+)-glyceraldehyde according to the conventions used for carbohydrates. That is, the configuration of the *highest numbered dissymmetric carbon atom* (penultimate carbon of a carbohydrate) may be directly correlated with the configuration of dextrorotatory glyceraldehyde. The symbol D_s is used to denote a compound that falls within the D configurational class—when correlated with D-(+)-serine according to the conventions used for amino acids. That is, the configuration of the *lowest numbered dissymmetric carbon atom* (α-carbon atom of an α-amino acid) may be directly correlated with the configuration of dextrorotatory serine. The assignment of the levorotatory tartaric acid discussed previously to the D_g and L_s configurational series appears in Figure III-24.

The threonine molecule is another example of a structure whose correlation with the D-(+)-glyceraldehyde configuration could potentially lead to confusion since natural (+)-threonine (Figure III-25) belongs to the L series of amino acids, but to the D series of carbohydrates. As indicated, levorotatory threonine could be classified as L_s-(−)-threonine to avoid any possible confusion.

Isocitric acid is a key biochemical intermediate whose structure is vitally involved in several discussions of biological stereospecificity in the following

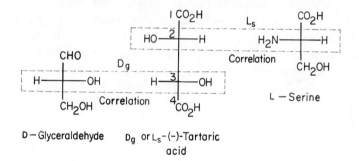

Figure III-24.   Assignment of (−)-tartaric acid to the D_g and L_s configurational series.

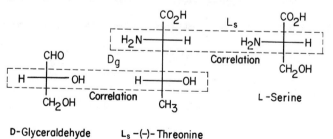

Figure III-25.   Correlation of (−)-threonine with D-glyceraldehyde and with L-serine.

chapters. The structures of the isocitric acid stereoisomers illustrate some further complications of the Fischer-Rosanoff configurational system. The absolute configuration of the particular dextrorotatory isocitric acid stereo-isomer involved in TCA metabolism has been established.[10,11] This natural isocitric acid is commonly referred to as D_s-(+)-isocitric acid since, as illustrated in Figure III-26, the configuration at the C-2 carbon may be directly correlated with the D-serine configuration. The isocitric acid structure possesses two asymmetric carbon atoms (C-2 and C-3), however, and it exists in the four stereoisomeric forms represented in Figure III-26. The D_s configurational nomenclature definitively specifies the stereochemistry only at the C-2 center of isocitric acid. It does not convey any stereochemical information regarding the configuration at C-3. The more complete designation *threo*-D_s-(+)-isocitric acid must be used to totally define the absolute configuration of this stereoisomer. The *threo* designation then defines the configuration at C-3 relative to the configuration at C-2, which, in turn, is correlated with the absolute configurational standard.

Thus far the examples of this chapter have pointed up two deficiencies of the Fischer-Rosanoff system of designating absolute configurations. In molecules containing multiple centers of dissymmetry, the D or L designation alone cannot communicate the total stereochemical information required in order to construct an accurate three-dimensional model. Furthermore, modifications of the basic system are necessary to define which dissymmetric center has been correlated with the original standard configuration. All systems of configurational designation based on correlations with a defined standard suffer from yet another serious deficiency. As the dissymmetric center to be defined becomes chemically unlike the original standard molecule, the correlation procedure becomes quite arbitrary.

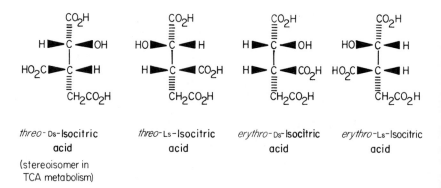

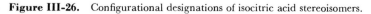

Figure III-26.   Configurational designations of isocitric acid stereoisomers.

Figure III-27.   1-Deuterioethanol (*R* absolute configuration shown).

In discussions of biological stereospecificity, for example, we must designate the absolute configurations present in molecules such as the 1-deuterioethanol stereoisomer pictured in Figure III-27. Correlation of the asymmetric center (C-1) of this alcohol with the standard D-(+)-glyceraldehyde or D-(+)-serine asymmetric centers would require a set of specific rules defining the corresponding chemical groups. Clearly, the "correct" application of the Fischer-Rosanoff configurational designations to such new dissymmetric groupings would require special knowledge of specific conventions in addition to a basic familiarity with this method of stereochemical nomenclature.

Let us summarize the foregoing discussions of the Fischer-Rosanoff method of designating configurational nomenclature: the method is most satisfactory for classifying structures within a defined series (e.g., the carbohydrates or the α-amino acids) and where a correlation with the D-(+)-glyceraldehyde substitution pattern may be made in a reasonably direct fashion. It is not a generally satisfactory method of designating configurations, however, since a knowledge of the correct three-dimensional structure will not immediately establish whether the configuration of a particular asymmetric center should be assigned to the D or L series. In addition, assignment of a compound to the D or L series will not automatically permit the *total* three-dimensional model to be constructed, even within a carefully defined chemical series.

CAHN-INGOLD-PRELOG ABSOLUTE CONFIGURATIONAL DESIGNATIONS: THE *R* AND *S* SYSTEM

Fortunately, an alternative method of specifying the absolute configuration of molecules has been developed. This system is variously called the Cahn-Ingold-Prelog method, after the scientists who have formulated it; the sequence method, referring to a key step in the application of the method; or the *R* and *S* system, since these letters are used to specify the absolute configurations determined by the method. Instead of correlating the configuration of a dissymmetric molecule with an arbitrary standard configuration, the sequence method assigns a priority to all possible functional groups. The absolute order of arrangement of these functional groups about an asymmetric center, as indicated by the order of priorities, is then used to specify the absolute configuration. This method of specifying absolute configurations is thus directly related to the fundamental difference between enantiomeric molecules—opposite chiralities or configurations. It is, therefore, generally applicable and permits a definitive configurational assignment to be made on the basis of a known three-dimensional structure without recourse to conventions other than the general rules governing all applications of the method. Conversely, given the R or S configuration of any given molecular structure, the correct three-dimensional representation can be readily reproduced. The sequence rule method does have certain inadequacies—in particular, the *R or S configurational designation of a molecule may be reversed by a reaction sequence that does not involve an actual inversion at the asymmetric center.* However, the overall convenience, simplicity, and general applicability of the R and S designations are such that this method is rapidly replacing the D and L configurational designations except in specific areas where the Fischer-Rosanoff correlations are firmly established by custom. *

The first step in applying the sequence method to an asymmetric center is to establish the relative priorities of the groups attached to the asymmetric atom. Initially it might be feared that this would require consulting a cumbersome list of possible functional groups arranged according to arbitrary priorities. In fact, a few easily remembered rules suffice to establish the necessary priorities even in complex structures. The first rule is that *atoms* of higher atomic number take precedence over those of lower atomic number. Thus we have the following series (listed in order of decreasing priority):

$$I > Br > Cl > F > O > N > C > H$$

Lone pairs of electrons, which are stereochemically important in such dissymmetric molecules as amines and sulfoxides, are assigned the lowest possible priority. The second rule is that *isotopes* of higher atomic weight take precedence over those of lower atomic weight. Thus: $_1^3H > _1^2H > _1^1H$. These

---

* It is worthwhile to observe that the gradual replacement of the Fischer-Rosanoff system of configurational designations by the Cahn-Ingold-Prelog system is a logical consequence of the development of the field of stereochemistry. The previous descriptions have emphasized that the Fischer-Rosanoff system is a *correlation* system, which could have been based on any arbitrary standard. It was well suited, therefore, for *correlating* the *configurations* at the asymmetric centers of molecules before any of the *absolute* three-dimensional structures had been actually established. Based on the earlier correlation studies and key X-ray determinations, the absolute three-dimensional structures of many molecules are now known. Therefore, it is now feasible to assign configurational nomenclature to molecules based on the established absolute arrangement of groups. Since the Cahn-Ingold-Prelog system requires and imparts this specific structural information, this method of configurational designation has become increasingly practical.

two simple rules will readily enable us to establish the relative priorities of many substituent groups. For example, in the 1-deuterioethanol molecule of Figure III-27 we assign (in order of decreasing priority) —OH = 1 > —CH$_3$ =2 > D = 3 > H = 4 as the priority of the groups substituting the asymmetric carbon atom.

In molecules of biological interest, two or more of the atoms attached directly to the chiral center will generally be the same. In such cases we establish the priorities between two substituting groups by considering the second atoms attached, then the third atoms, then the fourth, and so on—until the *first point* of difference in attached *atoms* is reached. Rule 1 is then applied to establish the correct priorities. * By way of illustration, the chloromethyl (—CH$_2$Cl), hydroxymethyl (—CH$_2$OH), and methyl (—CH$_3$) groups are all assigned priorities between, for example, —OH and —H, since they all have a carbon as the atom directly attached to an asymmetric center. Application of rule 1 *within* this carbon substituent series then leads to the expanded order

$$—OH > —CH_2Cl > —CH_2OH > —CH_3 > —H$$

since, considering the second atoms, Cl > O > H.† In a similar fashion a tertiary alkyl substituent will always take precedence over a secondary alkyl group which, in turn, takes precedence over a primary alkyl group; because, considering the first point of difference among the second atoms, C$_3$ > C$_2$H > CH$_2$. The following list, in order of decreasing priorities, can be derived for a series of alkyl substituents by application of the rules just stated:

$$—C(CH_3)_3 > —CH(CH_3)CH_2CH_2 > —CH(CH_3)_2 >$$
$$—CH_2CH(CH_3)_2 > —CH_2(CH_2)_2CH_3 > —CH_2CH_3 > —CH_3$$

In molecules of biochemical interest, many substituents will be bonded with multiple (double or triple) bonds. When an atom is attached to another by a double (triple) bond, both atoms are considered to be duplicated (triplicated) before rule 1 (and then, if necessary, rule 2) is applied to determine priorities. For purposes of determining priorities, therefore,

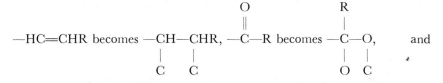

* It should be noted that, therefore, isotopic labeling, by itself, will only rarely change the *R* and *S* chirality designations of the nonisotopically labeled centers. Rule 2 (isotopes of higher molecular weight take precedence) is not applied until rule 1 has been applied to exhaustion.

† It should be noted that, at the first point of difference, a single atom of higher priority suffices to establish precedence; for example, —CH$_2$O > —CN$_3$.

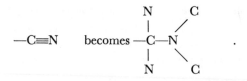

Only the unsaturated atoms are replicated, not the entire chain that is under consideration. A phenyl substituent (C_6H_5), therefore, becomes equivalent to

Application of this procedure leads to the following important ranking of common oxygenated carbon groups (in terms of decreasing priorities):

$$-CO_2CH_3 \left(\begin{array}{c} O-C \\ -C-O-CH_3 \\ O \end{array}\right) > -CO_2H \left(\begin{array}{c} O-C \\ -C-OH \\ O \end{array}\right) >$$

$$-CONH_2 \left(\begin{array}{c} O-C \\ C-NH_2 \\ O \end{array}\right) > -COCH_3 \left(\begin{array}{c} O-C \\ C-CH_3 \\ O \end{array}\right) >$$

$$-CHO \left(\begin{array}{c} O-C \\ -C-H \\ O \end{array}\right) > -CH_2OH$$

The cyano substituent —C≡N, with no oxygen attached, has lower priority than any of the above-mentioned oxygenated substituents, but it takes precedence over a saturated aminomethyl (—CH_2NH_2) substituent, which, in turn, takes precedence over the various hydrocarbon functional groups:

$$-C\equiv N \left(\begin{array}{c} N \\ -C-N<^C_C \\ N \end{array}\right) > -CH_2NH_2 > -C\equiv CH \left(\begin{array}{c} C\ C \\ -C-C-H \\ C\ C \end{array}\right) >$$

$$-CH=CH_2 \left(\begin{array}{c} C\ C \\ -C-CH_2 \\ H \end{array}\right) > -CH_2CH_3$$

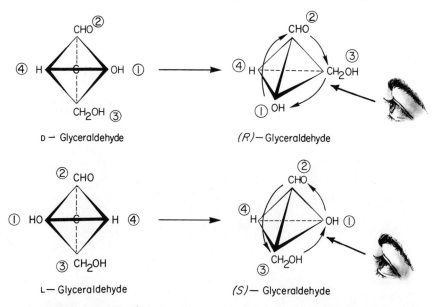

D — Glyceraldehyde   *(R)*— Glyceraldehyde

L— Glyceraldehyde   *(S)*— Glyceraldehyde

Figure III-28. Assignment of *R* and *S* configurational classifications to D and to L-glyceraldehyde.

and so on. In the manner illustrated by the series of rankings just derived, application of these procedures enables us to determine the relative priorities for any substituent found attached to an asymmetric center.*

The second step involved in specifying absolute configurations by the sequence method consists of orienting the tetrahedral molecule so that the lowest priority group is farthest from the viewing point. Finally, we count around the face of the tetrahedron containing the remaining three substituents *in order of decreasing priority*. A clockwise decreasing order (right = *rectus*) determines an *R* absolute configuration; a counterclockwise decreasing order (left = *sinister*) determines the *S* configuration. This is illustrated in Figure III-28 for D and L-glyceraldehyde. As Figure III-28 demonstrates, the sequence method of designating absolute configurations shows conclusively how enantiomers, having opposite chiralities, differ in the absolute order of arrangement of equivalent atoms in space.

* In certain complex branched structures it is necessary to consider the first difference found along "prior" branches, then along "secondary" branches, and so on. Illustration of the application of the sequence method to more complex compounds may be found in an introductory article by R. S. Cahn.[12] For a detailed description of the Cahn-Ingold-Prelog method of absolute configurational nomenclature, the articles in Ref. 13 should be consulted.

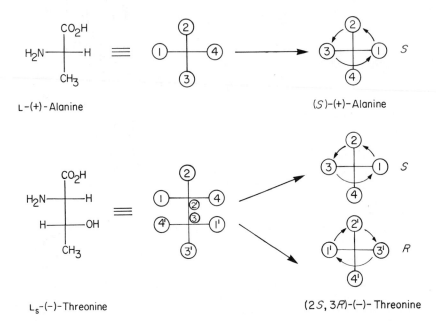

Figure III-29. Assignment of R and S configurational classifications to L-alanine and L$_s$-threonine.

Skill in using the R and S system of configurational nomenclature is largely dependent on practice in correctly viewing and/or manipulating three-dimensional models and perspective drawings. Beginning with a Fischer projection formula, the easiest procedure is to make use of the manipulation described in Figure III-17 in order to place the lowest priority group at the bottom of the projection. The R or S arrangement of the top three groups in terms of priority is then readily ascertained by the clockwise or counterclockwise numbering system. This is illustrated for L-(+)-alanine and for L$_s$-(−)-threonine in Figure III-29.

As we have just seen, the symbols R and S are written in parentheses, and then separated by a hyphen from the beginning of the correct chemical name. When two or more asymmetric centers are present within a molecule, the configuration at each center is determined separately and an appropriate locant number is added to the R or S designation. The individual configurational designations are then separated by commas, as illustrated for (2S,3R)-threonine. Racemic mixtures can be designated by RS; for example (RS)-glyceraldehyde and (RS,RS)-tartaric acid.

In Table III-1 the absolute configurations of several of the compounds which are drawn in perspective in this chapter are specified according to the sequence method. The reader should use this list to test his ability to properly assign the R and S configurations.

Table III-1   Some R and S Configurational Assignments

Figure	Absolute configurational assignment (Cahn-Ingold-Prelog)
III-2 (#1)	(1S,2S)-*trans*-1,2-Dichlorocyclopropane
III-2 (#3)	Troeger's base; S,S nitrogen configurations
III-3	(1S,2R)-*cis*-1,2-Dichlorocyclopropane
III-4	(1R,2R)-(d)-Tartaric acid
	(1S,2S)-(l)-Tartaric acid
	(R,S)-Tartaric acid (*meso*)
III-9	(2R,3S,4R,5R)-(+)-Glucose
	(2R,3S,4S,5R)-(+)-Galactose
III-16	(S)-(+)-Lactic acid
III-18	(3S,4R,5R)-Fructose
	(2R,3R,4R)-Ribose
but N.B.	(3S,4R)-2-Deoxyribose
III-19	(2R,3R)-Erythrose
	(2S,3R)-Threose
	(2S,3R)-(+)-*erythro*-3-Bromo-2-butanol
	(2S,3S)-(−)-*threo*-3-Bromo-2-butanol
III-23	(S)-(+)-Aspartic acid
III-26	(2R,3S)-Isocitric acid (*threo*-D$_s$-)
	(2S,3R)-Isocitric acid (*threo*-L$_s$-)
	(2R,3R)-Isocitric acid (*erythro*-D$_s$-)
	(2S,3S)-Isocitric acid (*erythro*-L$_s$-)
III-27	(R)-1-Deuterioethanol

PRO-CHIRAL CONFIGURATIONAL DESIGNATIONS: THE *PRO-R* AND *PRO-S* SYSTEM OF HANSON

For our particular purposes the R and S system of designating absolute configurations has a special advantage; it may be readily adapted to permit separate designations for the chemically like, but geometrically distinct paired groups of the nondissymmetric substrates defined in Tables II-2 and II-3. As we see in Chapter V, the relationship between a carbon atom bound

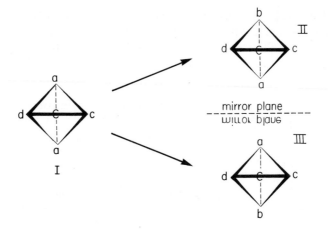

Figure III-30.   Conversion of a *pro*-chiral center into a chiral center by chemical substitution (cf. Table II-3).

to two equivalent and two nonequivalent groups (C_{aacd}) and biological specificity for chemically like, paired groups is analogous to the relationship between an asymmetric carbon atom (C_{abcd}) and optical activity. It should be apparent that the replacement of one of the chemically paired groups (a) of C_{aacd} by a fourth group (b; b $\neq$ a, c, or d) will convert this center into an asymmetric center: $C_{aacd} \rightarrow C_{abcd}$. Thus, by a single replacement, a nondissymmetric compound may be converted into a dissymmetric, chiral compound containing an asymmetric center. For this reason, the C_{aacd} carbon atom may be designated a *pro-chiral carbon center*, and the compound containing such a center may be designated a *pro-chiral molecule.* *

In Figure III-30, replacement of one of the paired substituents produces one dissymmetric molecule; replacement of the other member of the pair produces the enantiomer.

If the sequence of priorities of the substituting groups in Figure III-30 are d > c > b > a (decreasing priorities), then the structure **II** will be designated S and the enantiomeric structure **III** will possess the R configuration. Since the S enantiomer (**II**) arose by replacing the substituent at the top of the *pro*-chiral molecule **I**, this (a) substituent may be designated as the *pro-S* paired substituent. The (a) group at the bottom of the projection of molecule **I** is then designated as the *pro-R* substituent, since replacement of this group by (b) leads to a chiral molecule with the R configuration. *Furthermore, because replacement of one or the other of the paired (a, a) groups in this*

---

* The concept of *pro*-chirality and a proposed system of *pro-S, pro-R* configurational nomenclature was definitively developed by K. R. Hanson.[14]

example leads to enantiomeric molecules, the paired (a, a) groups of this particular example are defined as *enantiotopic paired substituents* (Table II-3). The formal rules for designating *pro-R* and *pro-S* groups may be stated as follows.

At a *pro*-chiral center, one of the paired substituents is arbitrarily selected and elevated in priority over the other paired substituent. The elevated priority is chosen so that the priorities *relative to the unpaired substituents remain unaltered* (i.e., in the previous example, $d > c > a = a \rightarrow d > c > b > a$). The configuration of the dissymmetric center (C_{abcd}) so derived is then designated according to the usual rules of the R and S system. If the derived center of dissymmetry possesses the S configuration, the paired substituent *originally elevated* in priority is designated the *pro-S* group; if the derived center possesses the R configuration, the elevated substituent is designated *pro-R*. Applying this system of stereochemical nomenclature, we can readily distinguish between the two nonequivalent hydrogens at the C-1 (*pro*-chiral carbon center) of ethanol. This is indicated in Figure III-31 by the subscripts R and S of the two nonequivalent (enantiotopic) paired hydrogen atoms. The statement that alcohol dehydrogenase from yeast (alcohol:NAD oxidoreductase) catalyzes the transfer of the *pro-R* hydrogen of ethanol to the *pro-R* position at C-4' of NADH thus becomes a concise, yet meaningful description of the biological stereospecificity of this enzyme (cf. pp. 159–161, 240–244).

Figure III-31.   Designation of *pro-R* and *pro-S* hydrogens at the *pro*-chiral center of ethanol.

Additional examples of the use of *pro*-chiral designations are given in Figure III-32 for (R)-glyceraldehyde, succinic acid, glycerol, and citric acid.

These examples illustrate several important aspects of *pro*-chiral centers and of *pro-R*, *pro-S* configurational designations. A detailed examination of the formal rule for designating between the paired substituents of a *pro*-chiral center leads to the observation that replacement of the various *pro-R* substituents will not always lead to an R product. For example, although isotopic substitution of any of the *pro-R* designated hydrogens in Figure III-31 by deuterium ($^2H > {}^1H$ in the R and S priorities) *will* lead to an R chiral center, replacement of these *pro-R* hydrogens by an amino group will lead in many cases to an S chiral center ($O > N > C$ in the R and S priorities). Similarly, the conversion of the carboxyl function on the *pro-R* branch of citric acid to an

$$\overset{\text{O}}{\underset{\|}{}} \qquad \overset{\text{O}}{\underset{\|}{}}$$

ester ($-C-OCH_3 > -C-OH$) will produce an R configuration at the

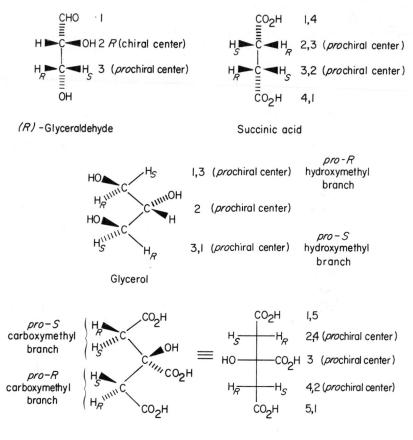

Figure III-32. Examples of *pro*-chiral configurational designations.

resulting C-3 chiral center, but conversion of the same carboxyl group to an amide function ($-\overset{\overset{\textstyle O}{\|}}{C}-OH > -\overset{\overset{\textstyle O}{\|}}{C}-NH_2$) will produce an S configuration. These potentially confusing observations arise, of course, from the necessarily arbitrary rules used to define the *pro-R* and *pro-S* positions. Although unfortunate, these observations do not decrease to a significant degree the usefulness of the *pro*-chiral designations.

A *pro*-chiral center C_{aacd} in which C_{cd} defines a σ plane of symmetry for the molecule—with the paired substituents (a) and (a) lying on opposite sides of the σ plane—has been termed a "*meso*-carbon atom."[15] A *meso*-carbon atom thus bears a close relationship to those "*meso*" compounds such as *meso*-tartaric acid, which possess σ planes. *Meso*-carbon atoms, by definition, possess reflective symmetry but lack rotational symmetry. The paired substituents of *meso*-carbon centers bear an enantiomeric (nonsuperimposable mirror image) relationship to each other and are therefore enantiotopic paired groups. The *pro*-chiral centers at C-1 of ethanol (Figure III-31), at C-2 of glycerol, and at C-3 of citric acid (Figure III-32), are *meso*-carbon atoms according to this definition; and the paired substituents at these centers are enantiotopic, chemically like, paired groups. The methylene groups at C-3 of (*R*)-glyceraldehyde, at C-2 of citric acid, or at C-1 and C-3 of glycerol (Figure III-32), however, can be contrasted with *meso*-carbon centers, since they do not fit into this especially defined class of *pro*-chiral centers. For example, it should be noted that replacement of the paired hydrogens at C-3 of (*R*)-glyceraldehyde leads to the diastereomeric (2*R*,3*S*) or (2*R*,3*R*)-glyceraldehyde molecules. Thus *the paired hydrogens at C-3 of (R)-glyceraldehyde possess a diastereomeric relationship in contrast to the enantiomeric relationships of the paired hydrogens at the C-1 carbon of ethanol.* The chemically like, paired groups at C-3 of (*R*)-glyceraldehyde, at C-2 of citric acid, and at C-1 and C-3 of glycerol are therefore diastereotopic paired groups (cf. Table II-3). We see in Chapter V that the distinction between diastereotopic paired groups and enantiotopic paired groups is important in differentiating among *types* of biological stereospecificities. Therefore, we will continue to distinguish between diastereotopic and enantiotopic, chemically like, paired groups. It does not seem particularly helpful, however, to divide the *pro*-chiral centers represented in Figures III-31 and III-32 into "*meso*-carbon *pro*-chiral centers" and "non-*meso*-carbon *pro*-chiral centers." It is important to recognize that the *pro*-chiral centers of ethanol and glyceraldehyde could be assigned to different classes, but here we refer to all such positions simply as *pro*-chiral centers. The differences among them are discussed in greater detail in Chapter V.

The *pro*-chiral designations applied to succinic acid in Figure III-32 illustrate some subtle aspects of the stereochemistry of this molecule. The succinic acid molecule is projected in a readily accessible conformation that possesses a C_2 axis of rotation. Since the C_2 rotation interchanges the two carboxyl groups and the two methylene carbons, positions 1 and 4 and positions 2 and 3 of succinic acid are totally equivalent paired positions and incapable of being differentiated (Table II-3). The *pro*-chiral designation, however, emphasizes that *two distinct pairs* of hydrogen atoms are present. The C_2 rotation will exchange H_S at C-2 with H_S at C-3 and H_R at C-2 with H_R at C-3. No possible rotation of the succinic acid molecule, however, will interchange any H_S atom with any H_R atom, while the other groups attain an equivalent conformation. Thus the H_S atoms at C-2 and C-3 are equivalent (superimposable by rotation), as are the H_R atoms at C-2 and C-3. The H_S and H_R pairs, however, are enantiotopic. Therefore they may be differentiated (Table II-3).

It is informative to contrast the stereochemical situation in succinic acid with that represented by citric acid and glycerol. Unlike succinic acid, the citric acid and the glycerol molecules lack rotational symmetry. Whereas the two arbitrary numbering schemes of chemical nomenclature that can be applied to succinic acid are equivalent ($C_1 = C_4$, $C_2 = C_3$) as indicated in Figure III-32, this is not the case in the glycerol and citric acid molecules. In one numbering scheme for citric acid, for example, the carboxyl group on the *pro-R* carboxymethyl branch would be called C-1; in the other numbering scheme, equally valid from the standpoint of chemical nomenclature, this stereochemically distinct carboxyl would be called C-5. Due to the absence of rotational symmetry in the citric acid molecule C-1 is not geometrically equivalent to C-5, and C-2 is not equivalent to C-4. Therefore, the *pro-R* and *pro-S* designations are a meaningful way of describing the true stereochemical situation prevailing in citric acid. We can describe the *pro-R* hydrogen of the *pro-R* carboxymethyl group as being stereochemically distinct both from its paired *pro-S* hydrogen *and* from the *pro-R* hydrogen of the *pro-S* carboxymethyl group. Describing a *pro-R* hydrogen at the C-2 position of citric acid could not convey this same stereochemical information. (But see the discussion of "*sn*" nomenclature below.) The *pro*-chiral designations that should be applied to the glycerol molecule to concisely but clearly define the stereochemical situation [i.e., H_R of the $(CH_2OH)_R$ group] are analogous.

Other systems of stereochemical nomenclature have been proposed which would convey the same factual information that is conveyed by the assigned *pro*-chiral designations in Figure III-32. We will illustrate an alternative system of stereochemical nomenclature for glycerol and its derivatives that has been proposed by the IUPAC-IUB Commission on Biochemical Notation.[16] Instead of using the *pro-R* and *pro-S* designation to distinguish the geometrically nonequivalent hydroxymethyl branches, the commission recommends a stereospecific numbering scheme for glycerol:

> The carbon atom that appears on top in that Fischer projection that shows a vertical carbon chain with the secondary hydroxyl group to the left is designated as C-1. To differentiate such numbering from conventional numbering conveying no steric information the prefix "*sn*" (for stereospecifically numbered) is used.

In this way the proposed rules for glycerol numbering would distinguish between the two numbering schemes for glycerol indicated in Figure III-32, so that only one numbering scheme (*sn*) would be valid. This proposed numbering system, then, explicitly acknowledges that when stereochemical considerations are included, the C-1 position of glycerol is not equivalent to the C-3 position. Applying this *sn* numbering scheme to the glycerol conformation of Figure III-32, the bottom or *pro-S* hydroxymethyl branch becomes the unique (*sn*)-glycerol-1 position.

It is clear that similar stereospecifically numbered schemes can be formulated to permit differentiation between the *stereochemically different* but *chemically arbitrary* numbering possibilities for citric acid illustrated in Figure III-32. For certain purposes, such an *sn* nomenclature system may be preferable; but for general discussions of biological stereospecificity, the *pro*-chiral notation in Figure III-32 conveys the pertinent information in an adequate and straightforward manner. Therefore, we use the *pro*-chiral notation in the subsequent discussions.

DIFFERENTIATION BETWEEN DISSYMMETRIC MOLECULES: CREATION OF DIASTEREOMERIC RELATIONSHIPS

The preceding description of the Cahn-Ingold-Prelog, or sequence method, of specifying absolute configurations should have resulted in a functional understanding of how dissymmetric molecules of opposite chirality differ by having a "clockwise" or a "counterclockwise" absolute arrangement of substituents in space. This functional comprehension of sequence nomenclature can be used to illustrate an additional fundamental aspect of dissymmetry. *Differentiation between inherently dissymmetric molecules depends on producing diastereomeric relationships and requires a dissymmetric environment.*

Using the sequence method of nomenclature we were able to designate the absolute configurations of dissymmetric molecules only by use of the dissymmetric concepts of "clockwise" and "counterclockwise." In the case of the Fischer-Rosanoff D,L designations, the absolute configurations could be assigned only by comparing the dissymmetric molecule in question with that of the D-(+)-glyceraldehyde standard. This standard thus served to create the dissymmetric environment necessary for distinguishing between opposite chiralities.

The recognition of this principle is of major importance in understanding biological stereospecificity. If an enzyme were nondissymmetric, then the interaction of substrate molecules of opposite chiralities (*R* or *S*) with the enzyme would be dissymmetric and therefore related as nonsuperimposable mirror images. Being dissymmetric, the interactions of our hypothetical nondissymmetric enzyme with the (*R*) substrate and with the (*S*) substrate could be *recognized* as *different* (related as nonsuperimposable mirror images), but they could be definitively *distinguished only* by recourse to an additional dissymmetric environment—a dissymmetric standard such as a right hand, a moving clock, or D-(+)-glyceraldehyde, or a dissymmetric medium such as plane-polarized light or a chiral solvent (see below). Like the enantiomeric molecules described earlier, the dissymmetric enzyme-substrate interactions would have exactly the same bond angles, bond lengths, and relative chemical groupings. Therefore, the energies of the enzyme-substrate interactions would be the same, and we would find experimentally that such important biochemical parameters as the Michaelis constant would be the same for both substrates. Similarly, the activation parameters for the reactions of the R substrate or the S substrate as catalyzed by our nondissymmetric enzyme would also be exactly the same—resulting in the same rate for either process. In short, a nondissymmetric enzyme could not produce the biological stereospecificity for chiral substrates that we customarily observe.

But, of course, enzyme structures *are* dissymmetric; moreover, only *one* of the possible enantiomeric forms of a given enzyme *will exist* in nature. The interaction of a single dissymmetric enzyme structure with the R or with the S form of a dissymmetric substrate will result in enzyme-substrate complexes that are *diastereomeric*, rather than enantiomeric. In contrast to the enantiomeric complexes that were related as nonsuperimposable mirror images, the diastereomeric enzyme-substrate possibilities will have different interatomic distances and interactions. Just as the different relative chemical groupings of the diastereomeric D-glucose and D-galactose molecules result in different chemical properties (D-glucose, monohydrate crystallized from water, mp 83°; D-galactose, monohydrate crystallized from water, mp 120°), the different relative chemical groupings of the diastereomeric enzyme-substrate complexes will result in different chemical properties, binding constants, activation parameters, and reaction rates.

The situation is analogous to that encountered during the chemical resolution of a racemic (RS) mixture. The enantiomeric molecules themselves can be recognized as different from models, but because of their equivalent chemical groupings they have exactly the same melting points, boiling points, solubilities, etc., and they cannot be chemically separated (differentiated). If, however, one forms salts between the racemic mixture and such dissymmetric reagents as *l*-brucine or *d*-camphorsulfonic acid, the resulting salts (e.g., D-substrate anion + *l*-brucine cation and L-substrate anion + *l*-brucine cation) are diastereomeric, not enantiomeric. Being diastereomeric, the salts will have different relative chemical groupings and thus different chemical properties. In particular, they will possess different solubilities and therefore can be chemically separated (differentiated) by fractional crystallization.

According to the principles just outlined, plane-polarized light will differentiate between chiral molecules only if plane-polarized light may be resolved into dissymmetric components whose interactions with molecules of opposite chiralities will be diastereomeric. A beam of light has time-dependent electric and magnetic fields associated with it. The variation in the electric field may be indicated by the vector E whose magnitude and direction are those of a sinusoidally oscillating field. If a light beam traveling in the z direction has the associated electric vector oscillations confined to the xz plane, the light is said to be plane-polarized. These relationships are illustrated in Figure III-33.

At this point it is difficult to see how the oscillating electric vector of our plane-polarized light can interact in a diastereomeric fashion with dissymmetric molecules of opposite chirality. Further analysis, however, indicates that the electric field vector of the plane-polarized light may be considered to be the resultant vector of the electric vectors from two in-phase *circularly-polarized* light beams of the same frequency. The electric field vector E_R of one circularly-polarized beam traces out a right-handed helix, and the electric field vector E_L of the second circularly-polarized beam traces out an enantiomeric left-handed helix. The projection onto the xy plane of the vectors E_R (associated with the right circularly-polarized beam) and E_L (associated with the left in-phase polarized beam) produces the previously described electric field vector E of the resultant plane-polarized light beam (see Figure III-34). This analysis makes it clear how plane-polarized light

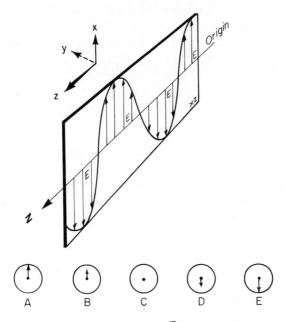

Time dependent projection of E onto xy plane.

Figure III-33. Representation of E associated with plane-polarized light.

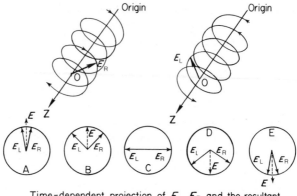

Time-dependent projection of E_L, E_R and the resultant
E onto the xy plane.

Figure III-34. Plane-polarized light represented as the resultant of two in-phase beams of right (E_R) and left (E_L) circularly-polarized light.

differentiates between enantiomeric molecules. The interaction of the right circularly-polarized beam with a molecule of right-handed chirality will be diastereomeric with respect to the interaction of the associated left circularly-polarized light with the same chiral molecule. The physicochemical manifestations of such diastereomeric interactions are distinct, just as the chemical properties of diastereomeric molecular structures are distinct. If the distinction between such polarized light-dissymmetric molecule interactions is analyzed in terms of the observable difference in refractive indices between the two circularly-polarized beams, the phenomenon is termed *circular birefringence*. The differences in refractive indices of two circularly-polarized beams interacting with a particular dissymmetric medium corresponds to the slowing of one beam relative to the enantiomeric beam. This causes a rotation of the resultant E vector ($E_R + E_L$) defining the plane of incident plane-polarized light (optical rotation). If the phenomenon of optical activity is analyzed in terms of the differential absorption of the circularly-polarized components of plane-polarized light, it is termed *circular dichroism*. All manifestations of optical activity, however, are due to the diastereomeric relationships produced when the enantiomeric circularly-polarized components of plane-polarized light interact with a dissymmetric environment, in full accord with the fundamental principle permitting physical differentiation between dissymmetric relationships.

The realization that plane-polarized light may be resolved into enantiomeric circularly-polarized beams will aid in understanding another often confusing aspect of optical activity—the optical activity of solutions. It is generally easier to accept the fact that the plane of polarized light will be rotated upon passage through a *solid* dissymmetric environment, such as a hemihedral quartz crystal, than to accept the fact that *solutions* of dissymmetric molecules will also display optical activity. In a crystal it is clear that individual dissymmetric molecules can have a fixed orientation with respect to a beam of light passing through the crystal and that the individual interactions of the polarized beam with the dissymmetric elements will be additive. In a solution of dissymmetric molecules, on the other hand, the orientations will be totally random; therefore, no fixed geometrical relationship exists between the individual dissymmetric molecules of the solution and the passing beam of polarized light. In this case, it may seem that the interaction of a polarized beam with a molecule of one orientation should be exactly canceled by the interaction of the beam with a molecule of the opposite relative orientation; just as the interaction of polarized light with one *dissymmetric conformation* of a molecule is canceled by the interaction of the beam with the *mirror image conformation*. The two situations, however, are not analogous. In contrast to the interactions of a beam of polarized light with different conformational isomers of molecules, the interactions with dissymmetric molecules of differing relative orientations, but of a single absolute configuration, is not enantiomeric. To visualize this situation, consider a model system based on nuts and bolts. The circularly-polarized components of the polarized light beam can be represented by nuts with right-handed threads (E_R) and with the enantiomeric left-handed threads (E_L).*

The dissymmetric molecule in our example is represented by a bolt containing right-handed threads. Only the right-handed nut will interact favorably (fit) the bolt with right-handed threads. No matter which way the left-handed nut is turned on a right-handed bolt, without destroying the threads themselves, it is impossible to tighten or loosen the nut. This model thus corresponds closely to the actual situation involved in optical activity as described previously in that one of the circularly-polarized beams preferentially interacts during passage through a dissymmetric environment. However, we are more concerned here with illustrating the effect of various possible orientations of the dissymmetric molecules. Two extreme orientations of a dissymmetric molecule relative to the passing beam of polarized light are represented by the bolts directed toward the light source and away from the light source, as in Figure III-35. Passage of the light beam from

---

* Right-handed and left-handed threaded systems are enantiomeric. A right-handed nut is turned clockwise to tighten and counterclockwise to loosen on a bolt with right-handed threads. A left-handed nut is turned counterclockwise to tighten and clockwise to loosen on a bolt with left-handed threads. Most threaded systems are right-handed. The bolts that tighten the centrifuge head in a Sorvall centrifuge are an example of a left-handed threaded system which is familiar to many biochemists.

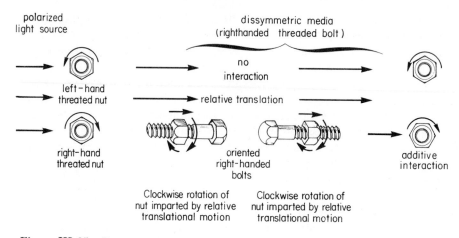

Figure III-35.   Representation of the additive interactions of circularly polarized components with randomly oriented dissymmetric structures.

left-to-right will correspond, therefore, to a left-to-right translational motion of the threaded nuts relative to the oriented bolts. In the interaction of the right-handed nut with the bolt pointing toward the light source, this clearly corresponds to a tightening or clockwise rotational interaction. In the case of the bolt pointing away from the source, this relative left-to-right translation will correspond to a loosening of the nut. This loosening requires a counter-clockwise rotation with respect to an observer on the right; *but to an observer at the left, at the light source, the rotation is again clockwise. The absolute rotational interaction of the right-handed nut with the bolt is thus the same for both extreme relative orientations of the dissymmetric right-handed bolt.* This illustrates how the interactions of a beam of polarized light with dissymmetric molecules having totally random relative orientations will nevertheless be additive, resulting in optical activity.

Nuclear magnetic resonance spectroscopy can also be used to demonstrate the principle that differentiation between enantiomeric molecules is dependent on the production of diastereomeric relationships. In contrast to diastereotopic nuclei, such as the paired hydrogens at C-4' of the dihydro-nicotinamide ring of NADH and NMNH, which will possess different and therefore potentially resolvable chemical shifts, it is impossible to resolve the absorptions of enantiotopic paired groups using the standard methods of nmr spectroscopy. Since enantiomeric molecules and enantiotopic paired groups possess equivalent relative chemical environments, enantiomerically related nuclei will absorb at exactly the same nmr frequency—unless a diastereomeric relationship can be created.

If a *chiral solvent* is used for the nmr examination, a dissymmetric solvation environment will be produced. Under these special conditions, diastereomeric relationships will be created between the solvent molecules and each enantiomer. The situation is analogous to the reaction of a dissymmetric acid with a mixture of enantiomeric bases to produce diastereomeric salts. Instead of salts of potentially differentiable solubilities, however, the use of a chiral nmr solvent creates diastereomerically solvated chemical groups with potentially differentiable nmr chemical shifts. In appropriate chiral solvents it is possible to determine the optical purity and absolute configurations of a variety of enantiomeric compounds, including alcohols, α-hydroxy acids, amines, α-amino acids, and sulfoxides.[17] The nmr spectrum of methyl alaninate enriched in the L enantiomer and dissolved in (R)-$(-)$-2,2,2-trifluorophenylethanol is presented in Figure III-36. The diastereomerically solvated O–CH$_3$ and C–CH$_3$ groups of the methyl D and L-alaninate enantiomers are clearly resolved in this spectrum. In an achiral solvent, the nmr spectrum of methyl DL-alaninate has only a single O–CH$_3$ absorption peak and one doublet for the C–CH$_3$ absorption. The resolved spectra of the methyl D and L-alaninate molecules in Figure III-36 clearly demonstrates that the relative chemical equivalence of enantiomeric molecules is modified by a dissymmetric environment. Interactions within the dissymmetric environment produce diastereomeric relationships that are chemically and physically differentiable. This principle, which is manifest in the phenomena of optical activity, the resolution of racemates by diastereomeric salt crystal-

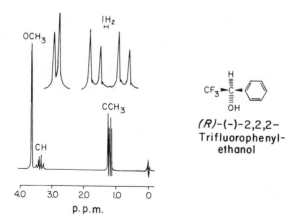

(R)-$(-)$-2,2,2-
Trifluorophenyl-
ethanol

Figure III-36. Portions of the 100-MHz nmr spectrum of partially resolved methyl L-alaninate in (R)-$(-)$-2,2,2-trifluorophenylethanol. The upper traces are expansions of the O–CH$_3$ and C–CH$_3$ absorptions.[17]

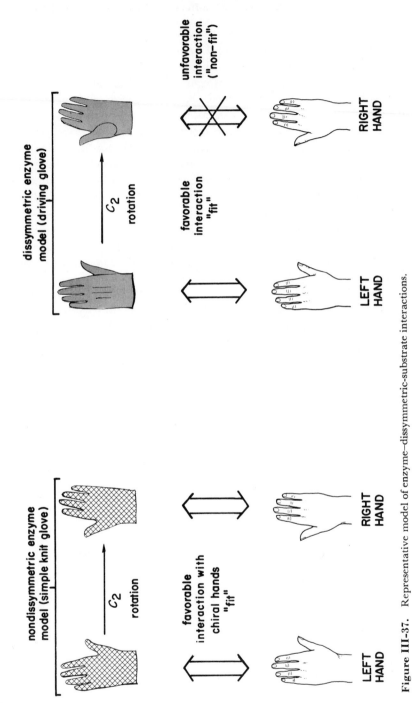

Figure III-37. Representative model of enzyme–dissymmetric-substrate interactions.

lization, and nmr spectra in chiral solvents, is also the basis of biological stereospecificity.

The glove-hand interactions pictured in Figure III-37 illustrate this fundamental principle governing discrimination of dissymmetric substrates and will serve as models of enzyme-substrate interactions. The dissymmetric substrate molecules of opposite chirality can be represented by the "chiral" relationship of right and left hands. A hypothetical nondissymmetric enzyme can be represented by the simple knit glove. Although the knit glove lacks rotational symmetry, it does possess a mirror plane. It is therefore nondissymmetric. We see that a nondissymmetric environment or reagent cannot discriminate between dissymmetric substances of opposite configurations because our nondissymmetric glove interacts equally well (fits) with a right or a left hand. If, however, we modify the glove by making it into a leather driving glove, we have a suitable dissymetric model for the unique stereochemical form of a normal dissymmetric enzyme. Now the dissymmetric glove fits only one absolute configuration of hand and easily permits differentiation between our "chiral" substrates.

In Chapter V, S. G. Cohen's models for the interaction of α-chymotrypsin with various substrates are outlined (Figures V-31–V-36). This particular enzyme may be viewed as a well-studied example, illustrating general features of the interactions of enzymatic catalytic sites with substrate molecules. The proposed interactions illustrated in Figures V-31–V-36 attempt to indicate how the different geometrical relationships found in enzyme-substrate interactions can account for the biological stereospecificity of α-chymotrypsin. In studying these examples of enzyme-substrate interactions, the principle governing differentiation of dissymmetries should be kept in mind. The representations of α-chymotrypsin-substrate interactions can account for the observed specificities because the pictured catalytic site is dissymmetric. Thus the interactions of the enzyme with chiral or with *pro*-chiral substrates will produce diastereomeric relationships between the catalytic groups and the reactive groups. These diastereomeric interactions, in turn, lead to differential chemical reactivity and the observed enzymatic stereospecificity.*

REFERENCES

1. "Researches on the Molecular Asymmetry of Natural Organic Products," L. Pasteur, *Alembic Club Reprints*, **14**, p. 38, University of Chicago Press, 1906.
2. R. Kuhn and T. Wagner-Jauregg, *Chem. Ber.*, **61**, 481 (1928).

* For a detailed analysis of possible chemical reaction types involving stereoisomers see, K. Mislow, *Introduction to Stereochemistry*, Section 3-3, W. A. Benjamin, 1966.

3. R. H. Sarma and N. O. Kaplan, *J. Biol. Chem.*, **244**, 771 (1969).

4. D. J. Patel, *Nature*, **221**, 1239 (1969).

5. R. H. Sarma, V. Ross, and N. O. Kaplan, *Biochemistry*, **7**, 3052 (1968).

6. M. A. Rosanoff, *J. Amer. Chem. Soc.*, **28**, 114 (1906).

7. J. M. Bijvoet, A. F. Peerdeman, and A. J. van Bommel, *Nature*, **168**, 271 (1951).

8. P. Brewster, E. D. Hughes, C. K. Ingold, and P. A. D. S. Rao, *Nature*, **166**, 178 (1950).

9. A. Wohl and F. Momber, *Chem. Ber.*, **50**, 455 (1917).

10. T. Kaneko and H. Katsura, *Chem. Ind. (London)*, 1188 (1960).

11. A. L. Patterson, C. K. Johnson, D. van der Helm, and J. A. Minkin, *J. Amer. Chem. Soc.*, **84**, 309 (1962).

12. R. S. Cahn, *J. Chem. Ed.*, **41**, 116 (1964).

13. R. S. Cahn, C. K. Ingold, and V. Prelog, *Experientia*, **12**, 81 (1956); R. S. Cahn, C. K. Ingold, and V. Prelog, *Angew. Chem. Int. Ed. Engl.*, **5**, 385 (1966).

14. K. R. Hanson, *J. Amer. Chem. Soc.*, **88**, 2731 (1966).

15. P. Schwartz and H. E. Carter, *Proc. Natl. Acad. Sci. (U.S.)*, **40**, 499 (1954).

16. IUPAC-IUB Commission, *J. Biol. Chem.*, **242**, 4845 (1967).

17. W. H. Pirkle and S. D. Beare, *J. Amer. Chem. Soc.*, **91**, 5150 (1969), and references therein.

Chapter IV

REPRESENTATIVE APPLICATIONS OF STEREOCHEMICAL ANALYSIS

In the three research projects from the published literature discussed in this chapter, the stereochemical problems were of paramount importance. These projects concretely illustrate several aspects of stereochemistry that have been developed in the first three chapters of this book. The discussion of these projects, therefore, will afford the student an opportunity to test his general understanding of stereochemistry.

DETERMINATION OF THE ABSOLUTE CONFIGURATIONS OF 3-HYDROXYPROLINES

The first project to be covered involves the identification of L-*cis* and L-*trans*-3-hydroxyproline by Sheehan and Whitney.[1] Hydrolysis of the antibiotic telomycin gave two amino acids which differed from any of the previously known α-amino acids, natural or synthetic. Based on the empirical formulas and certain chemical properties, whese novel amino acids were tentatively assigned the 3-hydroxyproline structure (Figure IV-1). Comparison of synthetically prepared samples of 3-hydroxyprolines and the

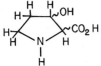

Figure IV-1. 3-Hydroxyproline. The wavy lines ($\sim$) indicate undetermined stereochemical assignments.

91

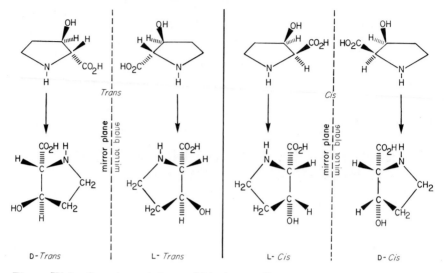

Figure IV-2. Stereoisomeric forms of 3-hydroxyproline.

amino acids obtained from telomycin indicated that the tentatively assigned structures were correct. The complete solution of this structural problem then involved only the determination of the absolute configuration at the two asymmetric centers of the 3-hydroxyproline isomers isolated from the antibiotic.

The four possible stereoisomers of the 3-hydroxyproline structure are depicted in Figure IV-2. Inspection of these forms reveals that no reflective symmetry elements are present; therefore, in contrast to the tartaric acids, the presence of two asymmetric centers will result in four distinct optical isomers. Since we are dealing here with an α-amino acid, by convention the absolute configuration at the α or C-2 position when compared to natural L-serine determines whether the structure belongs to the D or L series. As discussed previously, once the absolute configuration of one center in a molecule can be specified (here D or L), the absolute configurations at the remaining dissymmetric centers may be specified by indicating their configurations relative to the first center. In these structures we can indicate the absolute configuration of the hydroxyl group at C-3 by specifying whether it is *cis* (projecting on the same side of the five membered heterocyclic pyrrolidine ring) or *trans* (projecting on the opposite side of the pyrrolidine ring) to the carboxyl group at C-2. As illustrated in Figure IV-2, therefore, the designations L-*cis*, L-*trans*, D-*cis*, and D-*trans*-3-hydroxyproline totally specify the absolute three-dimensional structure of each stereoisomer. [The appropriate R and S configurational nomenclature would be $(2S,3R)$, $(2S,3S)$, $(2R,3S)$, and $(2R,3R)$, respectively].

It was determined that neither of the amino acids from telomycin was attacked by a preparation of hog kidney D-amino acid oxidase [D-Amino acid:oxygen oxidoreductase (deaminating)]. This enzyme has the normal expected biological stereospecificity for enantiomeric compounds, attacking, as the name implies, only those amino acids with the D configuration at the amino center. This experimental observation thus indicated that the two new amino acids were L-*cis* and L-*trans*-3-hydroxyproline. Since these two structures are not enantiomers, but diastereomers, they should have differing chemical properties. As anticipated, it was found that the two amino acids could be separated by chromatography or by electrophoresis. The entire structural problem now involved the establishment of whether one of the amino acids, say the "fast-moving" acid upon electrophoretic separation, was L-*cis* or L-*trans*.

There are a number of valid approaches to the solution of this final question. In general, we might have expected that such spectroscopic techniques as nmr or ir (infrared) could have provided the answer. In this case, however, the expected differences in the nmr (different coupling constants for *cis* and *trans* hydrogens at C-2 and C-3) and in the ir (intramolecular hydrogen bonding between the hydroxyl and carboxyl groups of the *cis* isomer) were insufficient to make an unambiguous structural assignment. Alternative approaches, such as forming a specific cyclic derivative dependent on a *cis* arrangement of the two functional groups, or a stereospecific synthesis of one isomer, would also have permitted a determination of which isomer possessed the *cis* relative stereochemistry and which the *trans*. In this case, however, Sheehan and Whitney combined both chemical and enzymatic reactions in an elegant degradative sequence that ultimately produced an optically active methoxysuccinamide molecule. The absolute configuration of this degradation product of the 3-hydroxyproline was then determined by direct comparison with a known methoxysuccinamide standard.

Since the supply of 3-hydroxyproline derived from the actual antibiotic was quite limited, the degradative sequence was applied to chemically synthesized materials. Owing to the nature of the synthetic reactions involved, the material actually synthesized was an isomeric mixture of 3-methoxy-prolines rather than of 3-hydroxyprolines. It was demonstrated, however, that treatment of the 3-methoxyproline to cleave the ether linkage occurred without changing the configuration of either chiral center. Thus 3-methoxy-proline of established stereochemistry could be converted into 3-hydroxy-proline of known stereochemistry for purposes of direct comparison with the isolated natural 3-hydroxyprolines. As expected in a general chemical reaction, the synthetic process was not completely stereoselective, and a mixture of the four possible 3-methoxyproline stereoisomers corresponding to the structures in Figure IV-2 was obtained. Fractional crystallization separated this mixture into two components. Since this type of fractionation depends on

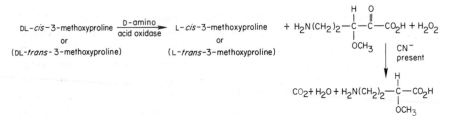

Figure IV-3. Action of D-amino acid oxidase on the racemic *cis* and *trans*-3-methoxy-prolines.

differing chemical properties, the method will separate diastereomeric, but not enantiomeric, molecules. In this case the separated diastereomeric fractions will be the DL-*cis* racemate and the DL-*trans* racemate (Figure IV-2). The enantiomeric molecules—for example, D-*cis* and L-*cis*-3-methoxyproline, present in one of the separated fractions—displayed identical chemical behavior and were not further separated by chromatography or electrophoresis. Sheehan and Whitney, however, took advantage of the principle that enantiomeric molecules can be differentiated when reacted with a dissymmetric reagent because the substrate-reagent interactions then produce chemically and physically distinct diastereomeric interactions. In this particular case, the separated *cis* and *trans* racemates were reacted with D-amino acid oxidase from hog kidney.

As previously mentioned, this enzyme preparation has a specificity for amino acids of D configuration. The D enantiomers present in each racemic mixture were therefore oxidized by the enzymatic reagent according to the equations in Figure IV-3.* The reactions in Figure IV-3 with the separated synthetic *cis* and *trans* racemates thus led to the production of L-*cis*-3-methoxyproline and L-*trans*-3-methoxyproline. Upon hydrobromic acid-catalyzed ether cleavage, one of these 3-methoxyproline samples gave a 3-hydroxyproline that was identical to the electrophoretic "fast-moving" isomer from telomycin; the other yielded a 3-hydroxyproline corresponding to the "slow-moving" isomer. The problem, therefore, was to determine which of the L-3-methoxyproline samples possessed the *cis* and which the *trans* relative stereochemistry.

This was determined by a chemical oxidation with permanganate. When treated separately with potassium permanganate, the L-3-methoxy-

---

* In the presence of cyanide, the catalase activity of the enzyme preparation was inhibited. This permitted the build up of hydrogen peroxide which reacted with the initial α-keto-acid product to produce carbon dioxide and the final acid product. In the absence of this second step, the amino acid oxidase would have been inhibited by the α-ketoacid initially produced.

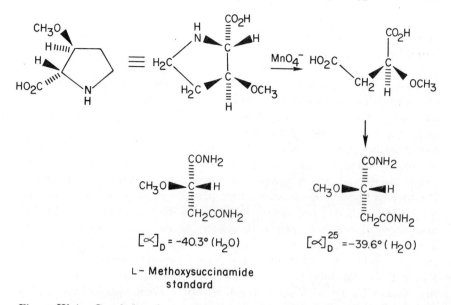

$[\alpha]_D = -40.3° (H_2O)$ $[\alpha]_D^{25} = -39.6° (H_2O)$

L - Methoxysuccinamide
standard

Figure IV-4.   Correlation of L-*trans*-3-methoxyproline with L-(−)-methoxysuccinamide.

prolines were converted to methoxysuccinic acids by oxidation of the two amino-substituted carbon atoms to carboxyl groups. Conversion of the methoxysuccinic acids to methoxysuccinamides permitted a direct comparison with methoxysuccinamide standards whose absolute configurations were known.[2] This permitted investigators to make the correct assignment of the *cis* and *trans* stereochemistry to the two L-3-methoxyproline samples and hence to the L-3-hydroxyprolines. These correlations are illustrated in Figure IV-4, where one of the possible L-3-methoxyproline structures has been arbitrarily chosen and its conversion to the known L-(−)-methoxysuccinamide shown. It can be seen that the production of L-(−)-methoxysuccinamide from our arbitrarily chosen L-3-methoxyproline sample requires that the sample be L-*trans*-3-methoxyproline. Since this L-3-methoxyproline sample is converted by hydrobromic acid into the "slow-moving" L-3-hydroxyproline isomer, the two new amino acids from telomycin are thus identified as L-*trans*-3-hydroxyproline (electrophoretic "slow-moving") and L-*cis*-3-hydroxyproline (electrophoretic "fast-moving").

A true understanding of the determination of absolute stereochemistry just described requires an appreciation for the necessity of the degradative step performed with D-amino acid oxidase. Although the *relative stereochemistries* at C-2 and C-3 of the *cis* and *trans* isomers of 3-hydroxyproline (or 3-methoxyproline) are obviously different, the *absolute stereochemistries* at C-3 of the D-*cis* and L-*trans* stereoisomers (and of the D-*trans* and L-*cis*) are the same. This situation arises, of course, because we are comparing the *relative* stereochemistry at C-3 and C-2 in D and L stereoisomers, where the *absolute* configuration at C-2 is reversed. This reversal is clearly indicated by the R and S configurational designations for the 3-hydroxyproline stereoisomers. Obviously, therefore, the isolation of L-($-$)-methoxysuccinamide from chemical degradation of a *trans*-3-methoxyproline will only permit stereochemical assignments to be made if the absolute configuration at C-2 of the original *trans*-3-methoxyproline is known. This was accomplished, in effect, by treating the separated *cis* and *trans* racemates with D-amino acid oxidase. The enzymatic degradation step would *not* have been necessary if Sheehan and Whitney had been able to carry out the permanganate oxidation steps upon derivatives of the L-3-hydroxyprolines isolated from the telomycin. In that case, conversion to a known methoxysuccinamide enantiomer would have permitted an unambiguous assignment of relative C-2 and C-3 stereochemistries. The student should carry out the correlation steps illustrated in Figure IV-4 with models and/or projection drawings for each of the four stereoisomeric forms of 3-methoxyproline until he is convinced of the validity of the above arguments.

ABSOLUTE CONFIGURATIONS OF THE *N*-METHYL-*N*-NEOPENTYL-4-METHYLCYCLOHEXYLAMINE OXIDE STEREOISOMERS

For our second example of research that involves stereochemical considerations, we examine an instance of knowledge of the reaction mechanism and the stereochemistry of the product permitting the absolute stereochemistry of the starting substrate to be assigned. In this study, Goldberg and Lam[3] synthesized the four optically active stereoisomers of *N*-methyl-*N*-neopentyl-4-methylcyclohexylamine oxide shown in Figure IV-5.

The synthesis of these stereoisomers began with the formation of a *cis*, *trans* mixture of *N*-neopentyl-4-methylcyclohexylamines. Since these *cis*, *trans* isomers are geometrical isomers possessing a diastereomeric (not enantiomeric) orientation of functional groups, they can be chemically differentiated. In this case, the mixture of stereoisomeric amines was separated into its *cis* and *trans* components by gas-liquid chromatography. The two isomeric amines were identified on the basis of differing nmr spectra. A *trans*-1,4-disubstituted cyclohexane will be largely frozen into a rigid diequitorial chair conformation (Figure IV-6). A *cis*-1,4-disubstituted cyclohexane, in contrast, may possess two equitorial-axial chair conformations of comparable energies (Figure IV-6). This difference in conformational possibilities resulted in observable differences in the cyclohexyl proton resonances that were used to identify the *cis* and *trans* isomers.

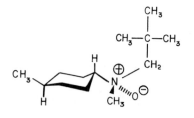

(S)— *trans*

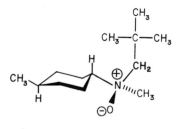

(R)— *trans*

S and *R* amine oxides with *trans* cyclohexane substitution.

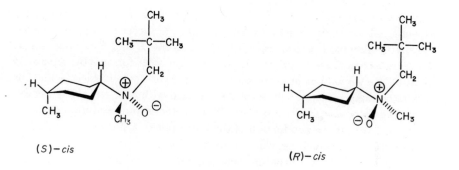

(S)—*cis*

(R)—*cis*

S and *R* amine oxides with *cis* cyclohexane substitution.

Figure IV-5.   Stereoisomers of *N*-methyl-*N*-neopentyl-4-methylcyclohexylamine oxide.

After isolation and identification of the *cis* and *trans* isomers they were treated separately with formic acid–formaldehyde to form the tertiary methyl amines and then with peracetic acid to produce the amine oxides.

In contrast to amines themselves, where "umbrella"-type inversions of the lone electron pair leads to racemization, amine oxides possess a nitrogen atom with a fixed tetrahedral geometry. Thus, from the pair of *cis* and *trans*-cyclohexylamines, the reaction sequence outlined in Figure IV-7 produced a pair of *cis* and *trans*-cyclohexylamine oxide racemates: (*RS*)-*cis*-*N*-methyl-*N*-neopentyl-4-methylcyclohexylamine oxide and (*RS*)-*trans*-*N*-methyl-*N*-neopentyl-4-methylcyclohexylamine oxide.

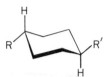

trans-1,4-diequitorial conformation

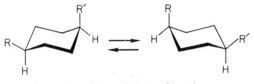

cis - 1,4-axial, equitorial conformations

Figure IV-6.   Conformational possibilities of 1,4-disubstituted cyclohexanes.

Since the (*RS*)-*cis* and (*RS*)-*trans*-amine oxide products are mixtures of enantiomeric stereoisomers, they must be separated (resolved) through the use of a dissymmetric reagent. Goldberg and Lam succeeded in resolving the racemic amine oxide mixtures into their respective *R* and *S* components by fractional crystallization of the diastereomeric salts formed by separately treating the *cis* and *trans* racemates with (−)-dibenzoyltartaric acid. The separated diastereomeric salts were then treated with base and the respective enantiomeric amine oxides reisolated. This synthesis and resolution sequence therefore leads to the production of the four optical isomers of *N*-methyl-*N*-neopentyl-4-methylcyclohexylamine oxide. The research problem now becomes analogous to the structure determination performed by Sheehan and Whitney, which was discussed previously. It had to be established which of the optically active amine oxides (+)-*trans*, (−)-*trans*, (+)-*cis*, and (−)-*cis* corresponded to the various structures projected in Figure IV-5.

It is perhaps worthwhile to indicate that, although the resolution of enantiomers by reaction with a dissymmetric reagent to produce separable diastereomeric products is always theoretically possible, experimentally the technique may fail. In this case, attempted resolutions through salt formation with both (−)-bromocamphorsulfonic acid and (−)-menthoxyacetic acid failed before Goldberg and Lam succeeded in obtaining separate diastereomeric salts with (−)-dibenzoyltartaric acid. Even with the latter reagent, the separation gave products of undetermined enantiomeric purity.*

* Cf. R. B. Woodward, M. P. Cava, W. D. Ollis, A. Hunger, H. U. Daeniker, and K. Schenker's comments on the resolution of enantiomers via diastereomeric salts, *Tetrahedron*, **19**, 259 (1963).

Figure IV-7. Formation of the *(RS)-trans-N*-methyl-*N*-neopentyl-4-methylcyclohexylamine oxides.

The four optically active amine oxides were subjected to the pyrolytic Cope reaction, and the methylcyclohexene produced in each case isolated and analyzed. Mechanistic studies of olefin formation by tertiary amine oxide pyrolysis, principally by Cope and co-workers,[4] have determined that this is a mild and stereospecific *syn* elimination (elimination of *cis* substituents) that proceeds through an essentially planar five-membered ring. The expected reaction transition state is illustrated in Figure IV-8 for the (+)-*trans* and (−)-*trans*-amine oxides.

Goldberg and Lam obtained the following results:

Amine Oxide Pyrolyzed	Predominant 4-Methylcyclohexene Formed
(−)-*trans*	dextrorotatory = (*R*)
(+)-*trans*	levorotatory = (*S*)
(−)-*cis*	levorotatory = (*S*)
(+)-*cis*	dextrorotatory = (*R*)

Since the absolute configurations of the 4-methylcyclohexenes were known,[5] these experimental results permitted the investigators to assign the absolute configurations of the starting amine oxides. As illustrated in Figure IV-8,

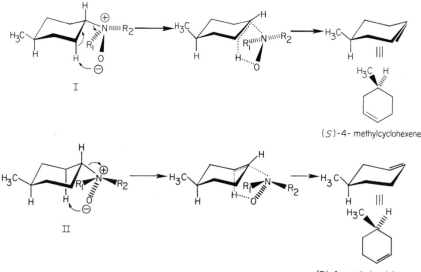

Figure IV-8. Formation of 4-methylcyclohexenes from the (±)-*trans*-*N*-methyl-*N*-neopentyl-4-methylcyclohexylamine oxide enantiomers.

when the amine oxide is placed in the proper orientation for a relatively planar five-membered transition state leading to *syn* elimination, one of the two acyclic alkyl substituents attached to the nitrogen is directed toward the cyclohexane ring, and the other is directed away from the ring. The neopentyl group is significantly bulkier than methyl. Steric considerations, therefore, dictate that the less crowded, more stable transition state will have methyl directed toward the cyclohexane ring and neopentyl directed away from the ring. This corresponds to the situations in Figure IV-8 if R_1 = methyl and R_2 = neopentyl. Structure **I** is therefore (*R*)-*N*-methyl-*N*-neopentyl-4-methylcyclohexylamine oxide and structure **II** is the (*S*)-enantiomer, according to Cahn-Ingold-Prelog configuration nomenclature.

The tabulated experimental results show that the (+)-*trans*-amine oxide isomer gave predominantly the (*S*)-(−)-4-methylcyclohexene. This dextrorotatory *trans*-amine oxide must therefore correspond to structure **I** of Figure IV-8, the *R* enantiomer. The *R* enantiomer, as illustrated, would be expected to produce a preponderance of the (*S*)-(−)-4-methylcyclohexene. Structure **II**, the *S* enantiomer where the less crowded transition state leads to the (*R*)-(+)-4-methylcyclohexene, was then assigned to the (−)-*trans*-amine oxide.

The validity of the foregoing stereochemical assignments are emphasized by the projections in Figure IV-9. Here the (*R*)-*trans* isomer (structure **I**) is rewritten in the more hindered, less stable, geometry necessary to produce (*R*)-(+)-4-methylcyclohexene and the (*S*)-*trans*-amine oxide (structure **II**) is rewritten in a conformation that would lead to the (*S*)-(−)-4-methylcyclohexene.

The student should use his own models and/or projection formulas to convince himself that the (−)-*cis*-amine oxide that produces a predominantly levorotatory (*S*)-4-methylcyclohexene should be assigned the *S* configuration and that the (+)-*cis*-amine oxide, producing predominantly the (*R*)-(+)-4-methylcyclohexene, should be assigned the *R* configuration.

The approach illustrated by the research of Goldberg and Lam—the assignment of the absolute configuration to a substrate from the known transition state geometry and the product stereochemistry—is the converse of one approach used to study enzymatic mechanisms. Quite often information regarding the geometry of the transition state of an enzymatic reaction can be inferred from the known configuration of the preferred enzymatic substrate and the product stereochemistry. As a special case of this approach to the study of enzymatic mechanisms, we can determine the absolute stereochemistry of a specific inhibitor molecule. This is one of the research aims of Kun and co-workers, whose work on the nature of the fluorocitrate inhibition of the aconitase reaction is presented as a third example of research involving stereochemical considerations.

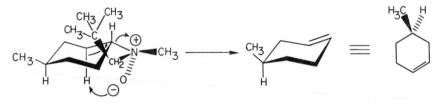

*(R)-trans-*amine oxide   *(R)* – 4 - methylcyclohexene

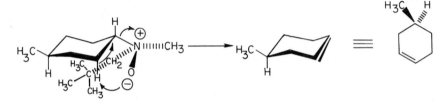

(S)-trans- amine oxide   *(S)* – 4- methylcyclohexene

Figure IV-9. Alternative transition state geometries (less stable) for the formation of 4-methylcyclohexenes from the *trans-N*-methyl-*N*-neopentyl-4-methylcyclohexylamine oxide enantiomers.

ABSOLUTE CONFIGURATION OF THE INHIBITORY MONOFLUOROCITRATE STEREOISOMER

Fluoroacetate has long been recognized as a metabolic poison. The material has been used as a rodenticide and its presence in the South African plant, *Dichapentalum cymosum* (Gifblaar), accounts for the toxicity of this plant toward livestock. In contrast to iodoacetate and its derivatives, which act as fairly general enzyme inhibitors by alkylating protein sulfhydryl groups, the antimetabolic activity of fluoroacetate is quite specific. Barron and co-workers found that the toxicity of fluoroacetate in animals, yeast, and bacteria was due to the inhibition of oxidative acetate metabolism.[6,7] Subsequent investigations have shown that the principal effect is blockage of the TCA cycle, by an inhibition of the aconitase [citrate (isocitrate) hydro-lyase] reaction by fluorocitrate, formed from the fluoroacetate.

The structure of the citric acids illustrated in Figure IV-10 should make it apparent that there are four possible stereoisomeric structures for mono-fluorocitric acid. Fluorine substitution could occur on the methylene carbon of either the *pro-R* or *pro-S* carboxymethyl branches of citric acid. In addi-tion, since each of the four methylene hydrogen atoms is stereochemically unique, a fluorine atom could replace either the *pro-R* or *pro-S* hydrogen

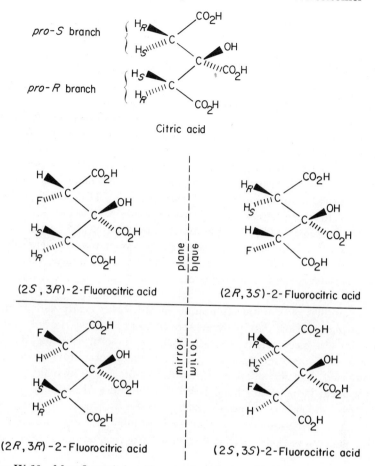

Figure IV-10. Monofluorocitric acids; enantiomeric monofluorocitric acids are separated by dashed lines; diastereomeric structures are separated by solid lines.

atoms on either of these two carboxymethyl branches. A nonstereoselective chemical synthesis will therefore produce a mixture of four monofluorocitric acid stereoisomers. It is apparent from Figure IV-10 (and from the *pro*-chiral nomenclature) that the two paired carboxymethyl branches attached to the central C-3 carbon of citric acid bear on enantiotopic relationship. It should also be apparent (from the structure and from the nomenclature) that the *pro*-S substituent at the methylene group of the *pro*-S carboxymethyl branch is *enantiotopic* to the *pro*-R substituent of the methylene group of the *pro*-R branch and *diastereotopic* to the *pro*-S substituent of the *pro*-R branch. The four monofluorocitric acids will therefore constitute a set of two diastereomeric pairs, with each diastereomeric pair, in turn, being composed of a pair of enantiomeric stereoisomers. After this review of the stereochemical possibilities for monofluorocitric acid, we are ready to discuss the stereochemical requirements for the fluorocitrate inhibition of aconitase.

The stereospecificity of the TCA cycle reactions between the carboxyl groups derived from acetate and those derived from oxaloacetate (historically referred to as the Ogston effect, cf. p. 12 and Figure I-3) is partially based on a stereospecific citrate synthetase reaction. Hanson and Rose have determined that the citrate formed by citrate synthetase derives its *pro-S* carboxymethyl branch from the acetate substrate, whereas oxaloacetate provides the *pro-R* carboxymethyl branch.[8] (This manifestation of biological stereospecificity is described in detail in Chapter V, pp. 151–154). The monofluorocitrate formed by citrate synthetase from fluoroacetate (e.g., in livestock grazing on Gifblaar) therefore will be (3R)-monofluorocitric acid.* Although this "natural" (3R)-monofluorocitrate isomer was known to be a potent inhibitor of aconitase, Speyer and Dickman had predicted that the fluorocitrate formed by the synthetase enzyme from fluoroxaloacetate and acetate would be an even more potent inhibitor of aconitase.[9]

The studies of aconitase inhibition by stereochemically defined monofluorocitrates carried out by Kun and co-workers have proved that this prediction was incorrect. In a series of papers, these investigators described the preparation and separation of the four monofluorocitrate stereoisomers and reported that only the (−)-*erythro* stereoisomer inhibited the aconitase activity of rat kidney mitochondria.[10-12] Furthermore, they found that this (−)-*erythro* isomer was stereospecifically formed by citrate synthetase from fluoroacetate and oxaloacetate. The noninhibitory (+)-*erythro*-monofluorocitrate isomer was stereospecifically formed from acetate and 2-fluoroxaloacetate.

It should be recalled that the prefixes *erythro* and *threo* are occasionally used to differentiate stereoisomers in which the functional groups have diastereomeric relationships analogous to those encountered in the diastereomeric aldotetroses, erythrose and threose (cf. Figure III-19). In *erythro* configurations, two like chemical groups may be simultaneously eclipsing. The *erythro* and *threo* configurations for monofluorocitric acids with the fluorine substitution on the R branch are displayed in Figure IV-11. It can be seen

---

* This is true even though the established stereospecificity of citrate synthetase will result in the replacement of one of the hydrogen atoms of the *pro-S* carboxymethyl branch of the citric acid molecule with a fluorine atom (cf., Figure IV-10). Substitution of one of these hydrogens by fluorine results in a (3R)-2-fluorocitric acid because —CH(F)(C) has a *higher* priority than —CO₂H in the Cahn-Ingold-Prelog system. In contrast, the tritiated citric acid formed by citrate synthetase from tritiated acetate will be (3S)-2-tritiocitric acid because —CH(^{3}H)(C) has a *lower* priority than —CO₂H.

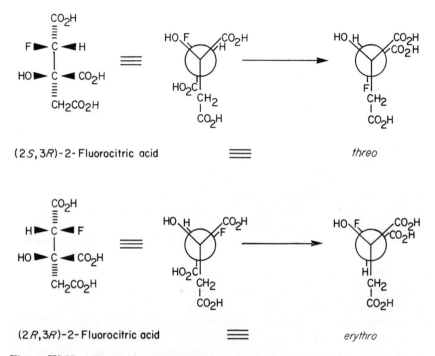

(2S,3R)-2-Fluorocitric acid   ≡   threo

(2R,3R)-2-Fluorocitric acid   ≡   erythro

Figure IV-11.   *Threo* and *erythro* configurations of the (3R)-2-fluorocitric acids.

from the Newman projections that this particular *erythro* stereoisomer corresponds to (2R,3R)-2-fluorocitric acid. The enantiomeric fluorocitric acid in which substitution has occurred in the *S* branch, (2S,3S)-2-fluorocitric acid, will also possess an *erythro* configuration. Indeed, a comparison of Figures IV-10 and IV-11 emphasizes that the *erythro* and *threo* configurational nomenclature designates diastereomers.

Kun found it possible to separate the *erythro* and *threo* racemates from a synthetic mixture of the four fluorocitrates by taking advantage of the chemical differences between the diastereomers. Although silica gel column chromatographic separations failed, paper electrophoresis separated the synthetic fluorocitrates into two components. In addition to the different electrophoretic properties, the different ir spectra, the different melting points, and the different pK's of the two isomeric components clearly established their diastereomeric nature. Later separations have been based on yet another observable chemical distinction—a difference in the solubilities of the cyclohexylammonium salts of the diastereomeric diethyl fluorocitrates. The observed stereoselectivity of the chemical synthesis (a 4:1 mixture of diastereomers was obtained) and the experimentally determined pK differences permitted the *erythro* and *threo* configurations of the separated racemates to be tentatively assigned.

Following the development of the appropriate separation and characterization techniques for fluorocitrates, Kun and co-workers were able to compare the chemical and biological properties of the biosynthetic fluorocitrates formed from fluoroacetate and oxaloacetate and from acetate and 2-fluoroxaloacetate. Electrophoretic analysis of these two biosynthetic monofluorocitrates showed that they *both* corresponded to the *erythro*-monofluorocitrate stereoisomers of the synthetic mixture. Biological assay, however, clearly revealed that the two biosynthetic acids were different. The biosynthetic *erythro*-fluorocitrate derived from fluoroacetate was a potent inhibitor of aconitase, but the *erythro*-fluorocitrate derived from fluoroxaloacetate did not inhibit aconitase.

As previously mentioned, the citrate synthetase system should produce fluorocitrate with the resulting fluorine atom on the R branch [(3R)-2-fluorocitrate] from fluoroacetate. If the enzymatic stereospecificity remains constant, then the fluorocitrate formed by citrate synthetase from 2-fluoroxaloacetate must be labeled on the S branch of citrate [(3S)-2-fluorocitrate]. Since all four methylene hydrogens of citrate are stereochemically unique, it follows that the two biosynthetic monofluorocitrates must be stereochemically different. Furthermore, since analytical electrophoresis revealed that both biosynthetic acids were *erythro* stereoisomers (no electrophoretic separation of the biosynthetic acids occurred), we know that these biosynthetic fluorocitric acids must be enantiomeric.

The biological assays reported by Kun et al. substantiate this stereochemical argument. The specific *erythro* stereoisomer formed by enzymatic synthesis from fluoroacetate was the most potent inhibitor of aconitase. The separated synthetic mixture of *erythro* stereoisomers, which would contain both ($+$) and ($-$)-*erythro* enantiomers, was half as potent an aconitase inhibitor on a weight basis. Finally, the *erythro* stereoisomer formed by enzymatic synthesis from 2-fluoroxaloacetate, which should be enantiomeric with the fluoroacetate-derived stereoisomer, was found to be inactive as an aconitase inhibitor. The relative level of aconitase inhibition shown by the unseparated synthetic mixture of fluorocitrates also indicated that the *threo* stereoisomeric forms lacked inhibitory activity.

The most recent paper by Dummel and Kun furnishes additional experimental evidence that the foregoing interpretation is correct. After separating the *erythro* and *threo* diastereomers from a synthetic mixture of monofluorocitrates by fractional crystallization of the cyclohexylammonium salts of diethyl fluorocitrate, the *erythro* racemate was further resolved into the ($-$) and ($+$) enantiomeric components. This was accomplished by first treating the diethyl *erythro*-fluorocitric acid racemate with (+)-deoxyephedrine. This led to a crystalline salt of (+)-deoxyephedrine and the diethyl (+)-*erythro*-fluorocitrate isomer. Subsequent treatment of the diethyl fluorocitrate recovered from the mother liquors with (−)-deoxyephedrine led to the isolation of a crystalline salt of (−)-deoxyephedrine and the diethyl (−)-*erythro*-fluorocitrate. The resolved (+) and (−)-*erythro*-fluorocitrate salts were then converted separately into the free diethyl fluorocitric acid forms, hydrolyzed to the tricarboxylic acids, purified as the barium salts, freed from the barium salt form, and finally crystallized from ethanol as the cyclohexylamine salts of (+) and (−)-*erythro*-fluorocitrate. Biological assays of the pure, resolved, *erythro*-fluorocitrates revealed that only the (−) enantiomer [based on the anticipated synthetase stereospecificity, this would be (2R,3R)-2-fluorocitrate] inhibited the activity of rat kidney mitochondria.

As described previously, originally the *erythro* stereochemistry of the inhibitory monofluorocitrate isomer was tentatively assigned on the basis of observed chemical differences between the separate *erythro* and *threo* racemates. A recent X-ray crystallographic analysis of the rubidium ammonium salt of the racemate of monofluorocitrate containing the inhibitory isomer indicates that this is indeed the *erythro* racemate [(2R,3R) and (2S,3S) racemate].[13] Although it remains to be conclusively established that the unique inhibitory *erythro* enantiomer is (2R,3R)-monofluorocitrate, the work of Kun et al. clearly indicates that only one stereoisomer of fluorocitrate inhibits the aconitase enzyme and that this specific isomer is uniquely formed in the citrate synthetase reaction between fluoroacetate and oxaloacetate.

Before leaving the problem of fluorocitric acid stereochemistry, it is worthwhile to discuss the different types of biological stereospecificity manifest in the citrate synthetase reaction with fluorinated substrates. 2-Fluor-

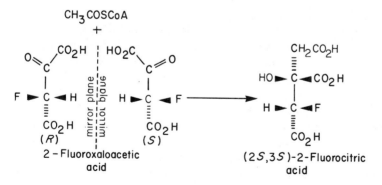

Figure IV-12. Stereoselective formation of (2S,3S)-2-fluorocitrate acid from (S)-2-fluoroxaloacetate in the citrate synthetase reaction.

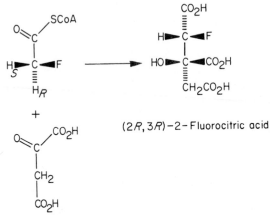

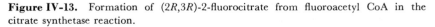

(2R,3R)-2-Fluorocitric acid

Figure IV-13. Formation of (2R,3R)-2-fluorocitrate from fluoroacetyl CoA in the citrate synthetase reaction.

oxaloacetate is a dissymmetric molecule possessing an asymmetric carbon at the C-2 position. Therefore, in addition to the previously mentioned stereospecificity of the citrate synthetase which leads to formation of the *pro-R* carboxymethyl branch of citrate from oxaloacetate, we should anticipate an enzymatic stereospecificity for a single enantiomer of the 2-fluoroxaloacetate. As illustrated in Figure IV-12, formation of the single (+)-*erythro*-fluorocitrate stereoisomer [(2S,3S)-2-fluorocitrate] from fluoroxaloacetate requires that the condensing enzyme specifically select the (S)-2-fluoroxaloacetate enantiomer for incorporation into fluorocitrate.

In the case of the enzymatic synthesis of fluorocitrate from fluoroacetate, the work of Kun et al. indicates that one stereoisomer results, the (−)-*erythro*-fluorocitrate [(2R,3R)-2-fluorocitrate]. Again, two types of biological stereospecificity are manifest in this result. The "normal" stereospecificity of the condensing enzyme leads to the formation of the *pro-S* branch of citric acid from acetate, and therefore, to the formation of (3R)-2-fluorocitrate from fluoroacetate. Placement of the fluorine atom stereospecifically in the 2R configuration, however, requires a second type of biological stereospecificity. In contrast to the reaction with 2-fluoroxaloacetate, this additional biological stereospecificity cannot be a discrimination between enantiomeric substrate molecules, since fluoroacetyl CoA exists in a single chiral form. Instead, substitution of one of the methyl hydrogen atoms of the acetate moiety with fluorine creates a *pro*-chiral center containing a pair of chemically like, diastereotopic paired hydrogens. This is emphasized in Figure IV-13 by designating the paired hydrogens *pro-R* and *pro-S*. Figure IV-13 illustrates that the citrate synthetase reaction in this case involves a stereospecific reaction at the *pro*-chiral center of the fluoroacetyl CoA. Stereospecific formation of the (2*R*,3*R*)-2-fluorocitrate could result either from the replacement of the *pro-R* hydrogen of the fluoroacetate unit with the carbon chain derived from oxaloacetate with retention of configuration at the acetate carbon, or from the replacement of the *pro-S* hydrogen of the fluoroacetate unit with inversion.* In Chapters V and VI we discuss in detail biological stereospecificity toward the chemically like, paired substituents at the *pro*-chiral centers of substrate molecules such as fluoroacetate.

REFERENCES

1. J. C. Sheehan and J. G. Whitney, *J. Amer. Chem. Soc.*, **85**, 3863 (1963).

2. T. Purdie and G. B. Neave, *J. Chem. Soc.*, 1517 (1910).

3. S. I. Goldberg and F.-L. Lam, *J. Amer. Chem. Soc.*, **91**, 5113 (1969).

4. A. C. Cope and N. A. LeBel, *J. Amer. Chem. Soc.*, **82**, 4656 (1960); A. C. Cope, E. Ciganek, C. F. Howell, and E. E. Schweizer, *ibid.*, **82**, 4663 (1960); A. C. Cope, N. A. LeBel, H.-H. Lee, and W. R. Moore, *ibid.*, **79**, 4720 (1957); A. C. Cope, C. L. Bumgardner, and E. E. Schweizer, *ibid.*, **79**, 4729 (1957).

5. S. I. Goldberg and F.-L. Lam, *J. Org. Chem.*, **31**, 240 (1966).

6. G. R. Bartlett and E. S. G. Barron, *J. Biol. Chem.*, **170**, 67 (1947).

7. G. Kalnitsky and E. S. G. Barron, *J. Biol. Chem.*, **170**, 83 (1947).

8. K. R. Hanson and I. A. Rose, *Proc. Natl. Acad. Sci. (U.S.)*, **50**, 981 (1963).

9. J. F. Speyer and S. R. Dickman, *J. Biol. Chem.*, **220**, 193 (1956).

10. D. W. Fanshier, L. K. Gottwald, and E. Kun, *J. Biol. Chem.*, **237**, 3588 (1962).

11. D. W. Fanshier, L. K. Gottwald, and E. Kun, *J. Biol. Chem.*, **239**, 425 (1964).

12. R. J. Dummel and E. Kun, *J. Biol. Chem.*, **244**, 2966 (1969).

13. H. L. Carrell, J. P. Glusker, J. J. Villafranca, A. S. Mildvan, R. J. Dummel, and E. Kun, *Science*, **170**, 1412 (1970).

* In Chapter VI recent experiments are described that establish that the citrate synthetase reaction involves a stereospecific inversion of configuration at the condensing methyl carbon (pp. 201–211).

Chapter V

BIOCHEMICAL DIFFERENTIATION BETWEEN CHEMICALLY LIKE, PAIRED SUBSTITUENTS

This chapter covers biological differentiation between the paired chemical groupings of a substrate molecule. This aspect of biochemical stereospecificity was introduced in Chapter I when the Ogston effect in TCA cycle metabolism was described. Then, in Figure I-5, the observed differentiation between the paired carboxymethyl branches of citrate in the aconitase reaction was rationalized in terms of Ogston's original proposal—a stereospecific three-point attachment between the substrate and the catalytic surface. The systematic development of such concepts as reflective and rotational molecular symmetries, diastereomeric and enantiomeric interactions, and chiral and *pro*-chiral configurational nomenclature in the intervening chapters now makes possible a more general and meaningful description of such manifestations of biological stereospecificity. In the analysis of this aspect of biological stereospecificity, we first examine the fundamental principles that determine the phenomenon. This will lead directly to the second important topic—how to predict which substrate molecules will be subject to such biological differentiations. Finally, in Chapter VI, we describe a series of representative examples in which the observation of such biological stereospecificity has led either to a more detailed knowledge of the enzymatic reactions or to a prediction of the type of enzymatic mechanisms involved.

EQUIVALENT AND NONEQUIVALENT CHEMICALLY LIKE, PAIRED GROUPS

As illustrated in Figure I-5, Ogston's postulation of a three-point attachment between substrate and enzyme permitted a pictorial rationalization of

110

the phenomenon of biological stereospecificity between paired chemical groups. Biochemical texts generally present comparable illustrations to explain the Ogston effect. These descriptions are helpful and permit us to visualize a possible differentiating process. Such descriptions, however, generally do not lead to an *understanding* of the phenomenon, because the principles underlying the differentiation process are not explicitly stated. The textbook explanations accompanying the illustrations also tend to stress the concept of a three-point substrate-enzyme interaction. This concept is derived from the original statements of Ogston that "it is possible that an asymmetric enzyme which attacks a symmetrical compound can distinguish between its identical groups, [providing] that sites a' and c' [on the enzyme surface, see Figure I-5] are catalytically different; [and] that three-point combination occurs between the symmetrical substrate and the enzyme."[1] Unfortunately, this emphasis on the concept of three-point combination between enzyme and substrate is misleading and promotes erroneous conclusions. Even the most wondrous of asymmetric enzymes cannot differentiate between paired groups that are chemically and geometrically equivalent— no matter how many combination points are involved.

Within the operations of the TCA cycle, for example, the label from an acetate-1-^{14}C molecule, which is stereospecifically confined to the *pro-S* branch of citrate, to the γ-carboxyl of isocitrate, and to the γ-carboxyl of α-ketoglutarate, subsequently becomes equally distributed between the two carboxyl groups of succinate (cf. Figure I-3). Even asymmetric enzymes, therefore, cannot distinguish between the truly equivalent carboxyl groups of the succinate molecule. The primary alcohol groups of glycerol and those of dihydroxyacetone (Figure V-1) furnish an additional common example of this type of contrasting biochemical behavior.

Ogston's original insight and proposal, at least as interpreted in most biochemical texts, does not provide a basis for understanding and predicting the contrasting biochemical behavior of succinate and citrate or of glycerol

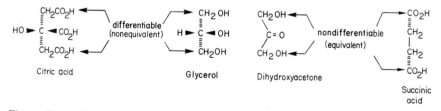

Figure V-1. Examples of biological substrates possessing differentiable and nondifferentiable chemically like, paired groups.

and dihydroxyacetone.* The analyses of biological stereospecificity between chemically like, paired groups that have been carried out by Wilcox,[2] Hanson,[3] Hirschmann,[4] and Schwartz and Carter,[5] have all advanced understanding of this phenomenon by stressing that the inherent structural features of substrate molecules make such biological discrimination possible. Interaction of the substrate molecule with an enzyme surface is therefore merely the mechanism that permits the inherently different groups of the substrate to be recognized, selected, and differentially reacted.

The analyses of biological stereospecificity carried out by the authors just named differ considerably in depth, but they may be divided into only two contrasting approaches. The analyses of Wilcox, Schwartz and Carter, and Hanson tend to stress the reflective symmetry relationships present within a substrate exhibiting stereospecific reactivities. Hirschmann's analysis, on the other hand, stresses the lack of rotational symmetries in these substrates. The two approaches complement each other and are equally valid. Since both these methods of analyzing the critical structural relationships can increase overall understanding of biological stereospecificity between chemically like, paired groups, both approaches are discussed in this chapter. Analysis of the reflective relationships present within potential biological substrates leads more directly to an understanding of the principles underlying biological stereospecificity between chemically like, paired groups. Therefore, this approach is described first.

DIFFERENTIATION BETWEEN ENANTIOTOPIC PAIRED GROUPS

According to the symmetry designations discussed and summarized in Chapter II, citric acid is a nondissymmetric molecule. Although the molecule lacks symmetry of the rotational class, it possesses reflective symmetry. In particular, it has a mirror plane of symmetry that passes through C-3 perpendicular to the five-carbon backbone of the molecule. The presence of this mirror plane, of course, means that citric acid may not exist in optically active, enantiomeric, forms. However, this mirror plane, which passes through

* A correct analysis and explanation of the phenomenon of biological stereospecificity between paired chemical groups should begin by emphasizing the structural features inherent in the substrate molecules. Only after analyzing the substrate to determine the types of *potentially differentiable* groups present, should we consider how interactions with an enzyme, such as Ogston's proposed three-point combination, will permit biological stereospecificity to become manifest. To be sure, if we return to the basic proposal of Ogston (the asymmetric three-point attachment) and visualize the appropriate interactions, the contrasting relationships illustrated in Figure V-1 are predicted. However, as described later, the same predictions are more readily derived from an appropriate analysis of the inherent structural features of the substrate molecules.

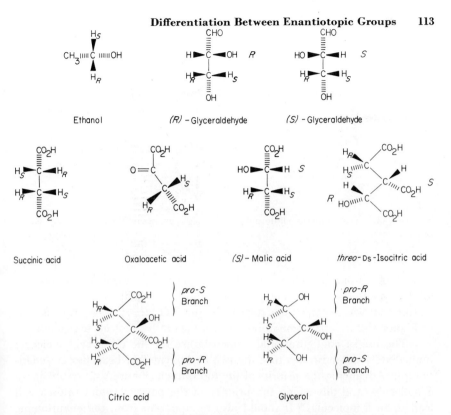

Ethanol   (R) – Glyceraldehyde   (S) – Glyceraldehyde

Succinic acid   Oxaloacetic acid   (S) – Malic acid   threo- D_S -Isocitric acid

Citric acid   Glycerol

Figure V-2. Chiral and *pro*-chiral designations for molecules illustrating various aspects of biological stereospecificity.

the C-3 atom, the C-3 hydroxyl, and C-3 carboxyl, divides the citric acid molecule into mirror image halves. Furthermore, the lack of all elements of rotational symmetry in the citric acid molecule means that these mirror image halves will be *enantiotopic*; that is, interchanged *only* by reflective symmetry operations. Schwartz and Carter, in their analysis of biological stereospecificity, proposed a special designation for a carbon atom such as the C-3 of citric acid, which is substituted with two chemically like, paired groups and two other chemically distinct groups (C_{aacd}). Since the C_{aacd} grouping possesses a mirror plane that passes through C, (c), and (d), and relates (a) and (a) as enantiotopic parts of the molecule, Schwartz and Carter proposed that a C_{aacd} center be called a *meso*-carbon atom to stress the analogy with such molecules as *meso*-tartaric acid. Discrimination between the chemically like, paired (a) and (a) groups of such *meso*-carbon atoms characterizes an extensive class of biological stereospecific reactions.

Actually, as pointed out in the more systematic analysis of Hanson, the

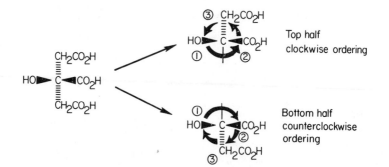

Figure V-3. Enantiomeric relationship (ordering) of groups on the mirror image halves of citric acid.

meso-carbon atom designated by Schwartz and Carter is only one important class of *pro*-chiral center. The *pro*-chiral and chiral configurational nomenclature that was discussed in detail in Chapter III (*pro-R, pro-S, R, S*) is used in Figure V-2 to label some important representative molecular structures.

The use of configurational nomenclature for the citric acid molecule immediately demonstrates that the two carboxymethyl branches are enantiotopic. Advancing the priority of the top branch (Figure V-2) results in an *S* molecule, and advancing the priority of the paired bottom branch will lead to an *R* molecule.* It should also be apparent from the enantiotopic relationship dictated by the *pro-S* and *pro-R* designations that *the absolute ordering of chemical groupings around C-3 of citric acid is opposite for the top and bottom halves of the molecule.* This is illustrated in Figure V-3 using the Cahn-Ingold-Prelog priority rules.

The basic principle permitting stereospecificity between chemically like, paired groups should now become apparent. *Differentiation between the two enantiotopic halves of the citric acid molecule is exactly analogous to the differentiation between separate enantiomeric molecules.* Enantiomeric molecules will interact with a dissymmetric agent (biological enzyme or chiral chemical resolving agent) to produce physically and chemically distinct diastereomeric combinations. The separate enantiotopic halves of the citric acid molecule will likewise interact with a given dissymmetric agent (biological enzyme or chiral chemical reagent) to produce physically and chemically distinct

---

* Unless the priority of the paired branch is advanced beyond the —CO_2H priority. The substitution of a hydrogen atom on the *pro-S* carboxymethyl branch of citric acid by fluorine to yield a (3*R*)-2-fluorocitric acid is such an example. The rules for *pro*-chiral designations in Chapter III (p. 77) should be carefully reexamined and compared with the fluorocitric acid nomenclature of Figure IV-10 to clarify this potentially confusing point.

diastereomeric combinations. In both classes of stereospecific reaction, the diastereomeric relationships of the two reaction possibilities will produce different steric and polar interactions and will result in reaction pathways of distinctly different energies. In this way the thermodynamics of the reactions of enantiomeric molecules, and of enantiotopic portions of a single molecule, become different, and preferential reactivities are observed. Again, as stressed in Chapter III, differentiation between enantiomeric possibilities is dependent on the creation of diastereomeric relationships.

The steric relationships at C-2 of glycerol are exactly analogous to those found at C-3 of citric acid (cf. Figure V-2). The biological stereospecificity manifest in the exclusive formation of L-α-glycerol phosphate in the glycerol kinase reaction, therefore, is exactly analogous to the stereospecificity shown by the aconitase reaction with citrate. Again, the combination of the *pro-S* hydroxymethyl branch of glycerol with a dissymmetric enzyme will be diastereomeric with respect to the combination of the *pro-R* branch of glycerol with this same dissymmetric enzyme. These diastereomeric interactions are thermodynamically different and the biological reaction specifically esterifies one enantiotopic portion of the molecule (the *pro-R* branch) to produce L-α-glycerol phosphate.

That the biological process in such reactions is differentiating between chemically like, but geometrically distinct paired groups was demonstrated in elegant fashion by Karnovsky et al.[6] DL-Serine-3-^{14}C was resolved, and each enantiomer deaminated stereospecifically to the respective D or L-glyceric acid-3-^{14}C. As illustrated in Figure V-4, L-serine-3-^{14}C yields the L (or S) glyceric acid-3-^{14}C, whereas D-serine yields the D (or R) glyceric acid-3-^{14}C. The two labeled glyceric acid enantiomers were then reduced to produce differentially labeled glycerol molecules. It can be readily demonstrated that one of the glycerol molecules is labeled in the *pro-R* hydroxymethyl branch and the other is labeled in the *pro-S* branch by this procedure. Mentally converting the ^{14}C-labeled primary alcohol center to an aldehyde group produces (R)-glyceraldehyde-1-^{14}C from one substrate and (S)-glyceraldehyde-1-^{14}C from the other substrate. Indeed, the two differentially labeled glycerol molecules can be designated (R)-glycerol-1-^{14}C and (S)-glycerol-1-^{14}C, as indicated. The R enantiomer could also be designated (R)-glycerol-3-^{14}C and the S isomer (S)-glycerol-3-^{14}C, but this merely reflects the ambiguity of nonstereochemical numbering schemes for molecules such as glycerol (cf. pp. 80–81). The two glycerol substrates are different because it is impossible to simultaneously superimpose the two ^{14}C-labeled groups and the remaining related chemical groups.

The two labeled glycerol substrates were incubated separately with liver slices and the glycerol phosphates resulting from the glycerol kinase reaction isolated. *Both substrates yielded exclusively the L-α-glycerol phosphate.*

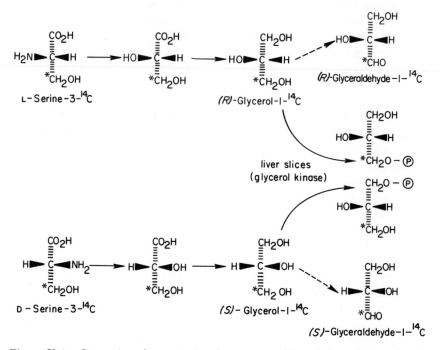

Figure V-4. Conversions demonstrating the stereospecificity of glycerol kinase for the *pro-R* branch of glycerol. [The position of the ^{14}C labels are indicated by asterisks. The phosphoryl group (PO_3H_2) is indicated by ⓟ.]

However, analysis of the position of the ^{14}C in the L-α-glycerol phosphate products revealed that the ^{14}C label from the (S)-glycerol-1-^{14}C substrate was present exclusively at the carbon bearing the unesterified primary alcohol group; the label from the (R)-glycerol-1-^{14}C substrate was located at the carbon bearing the phosphate ester group. This result, then, explicitly demonstrates that the stereospecific glycerol kinase reaction differentiates between the chemically like, but geometrically distinct hydroxymethyl branches of the glycerol substrate. *The experiment also demonstrates that the presence of the isotopic label is not involved in determining the biological stereospecificity and does not alter the specificity of the enzyme for esterification of the pro-R portion of the molecule.*

    This aspect of biological stereospecificity between chemically like, paired groups is occasionally a source of misunderstanding. Initially some students feel that the differentially labeled substrates used to *detect* this type of biological stereospecificity are, in themselves, the *source* of the observed stereospecific behavior. For example, the observation that the substrates used in the previous experiments could be designated (S)-glycerol-1-^{14}C and (R)-glycerol-1-^{14}C may lead to the misconception that this type of biological stereospecificity is identical to the more familiar biological differentiation between separate enantiomeric substrate molecules. Indeed, some of the earlier analyses of biological stereospecificity between chemically like, paired groups tended to give this impression.[7] In biological differentiation between enantiomeric molecules, however, only one enantiomer is reacted (cf. the reaction catalyzed by D-amino acid oxidase discussed in Chapter IV). In the experiments carried out by Karnovsky et al., in contrast, glycerol kinase catalyzes the reactions of *both* (R)-glycerol-1-^{14}C and (S)-glycerol-1-^{14}C. The biological stereospecificity between chemically like, paired groups *inherent in the glycerol kinase reaction becomes manifest by the production of differentially labeled L-α-glycerol phosphate products from these two substrates.* The isotopic labels, therefore, are only a means of detecting the stereospecificity between paired groups that is present in the catalyzed reactions of unlabeled substrates as well as of the labeled substrates.

This point may be emphasized by stressing that only a minor fraction of the molecules referred to as (R)-glycerol-1-^{14}C are in fact labeled with a $^{14}_{6}$C isotopic nucleus. In a substrate possessing a specific activity of 1 Ci/mol of $^{14}_{6}$C, for example, a consideration of Avogadro's number (6.02×10^{23}) and the half-life (5.77×10^3 years) indicates that only about 2% of the molecules are isotopically substituted with $^{14}_{6}$C. In the case of tritium substitution, the significantly shorter half-life (12.26 years) means that, in a substrate possessing a specific activity of 1 Ci/mol of $^{3}_{1}$H, only about 10^{-2}% of the molecules is isotopically substituted with $^{3}_{1}$H. Designating a substrate as (R)-glycerol-1-^{14}C or (R)-1-tritioglycerol does not reflect the actual chemical composition.

Determinations of the biological stereospecificity of a process, however, do not require that all the reacting molecules be isotopically labeled. These determinations will be valid so long as the *final stereochemical result* of the reaction of the isotopically labeled substrate molecules is the same as the *final stereochemical result* of the reaction of the bulk of the substrate molecules. Fortunately, isotopic substitution changes neither the geometry of a substrate molecule nor the inherent stereochemical features of enzyme-substrate interactions. Therefore, the final stereochemical result observed in the biological stereospecific reaction of an isotopically labeled substrate molecule *will* accurately reflect the stereochemical result of reaction of the unlabeled substrate molecules also present.*

* See, however, the more complex situation that pertains in determinations of stereospecific reactions of chiral acetate substrates; Chapter VI, pp. 199–201.

THE KINETIC ISOTOPE EFFECT

Although isotopic substitution *does not change the final stereochemical results* of a reaction, isotopic substitution *frequently does result in an appreciable change in the rate of a reaction*. This result of isotopic substitution is the *kinetic isotope effect* commonly involved in studies of chemical reaction mechanisms. *A kinetic isotope effect* is a difference in chemical reactivity arising from isotopic substitution in a reacting molecule. It occurs because isotopic atoms (e.g., $_1^1H$, $_1^2H$, $_1^3H$), although chemically alike, are physically distinct species. One manifestation of this physical difference between different chemical isotopes is a difference in nuclear stability. The $_1^1H$ and $_1^2H$ atoms have stable nuclei, but $_1^3H$ undergoes radioactive decay. Another manifestation of the difference is an alteration in the fundamental vibrational frequency of a particular bond. The vibrational frequency for the carbon–hydrogen bond stretch, for example, is inversely proportional to the square root of the rest mass of the carbon and hydrogen nuclei involved. Substitution of an $_1^1H$ atom by an $_1^2H$ atom with twice the nuclear mass will thus result in a shift of the vibrational resonance of this stretching mode to lower frequency. *

A relatively simple analysis shows how this difference in vibrational frequency can be correlated with reaction rate differences. If a chemical reaction involves breaking a carbon–hydrogen bond, the process can be viewed as a continuous stretching of this bond to the point at which the chemical bonding forces are overcome. In the transition state, therefore, the carbon–hydrogen vibration is converted into a translation of the hydrogen atom away from the carbon. Since the chemical bondings of $_1^1H$ and $_1^2H$ are fundamentally similar, the same energy maximum must be attained before the chemical bonding forces between a C–$_1^1H$ or a C–$_1^2H$ are overcome. Quantum mechanics, however, tells us that the lowest permissible vibrational energy state of the carbon–hydrogen bond is not located at the bottom of the potential energy well for this system, but lies above this minimum by the amount of the *zero-point energy*. This zero-point energy is equal to $\frac{1}{2} h\nu$, where h is Planck's constant and ν is the frequency of the bond stretching vibration. We have already seen how the difference in mass of the $_1^1H$ and $_1^2H$ isotopic nuclei will lead to a lower vibrational frequency for a C–$_1^2H$ stretch than for a C–$_1^1H$ stretch. Thus the zero-point energy of the C–$_1^2H$ vibration (ca. 3 kcal/mole) will be less than the zero-point energy of the C–$_1^1H$ vibration (ca. 4 kcal/mole).

In a reaction that involves breaking carbon–hydrogen bonds, therefore, it is necessary to raise the carbon–hydrogen vibrational energy from different ground state minima for C–$_1^1H$ and C–$_1^2H$ to the same transition state maxi-

---

* The observed position of the absorption for the C—$_1^1H$ stretch in chloroform is 2915 cm^{-1}; the C—$_1^2H$ stretch in deuteriochloroform occurs at 2256 cm^{-1}.

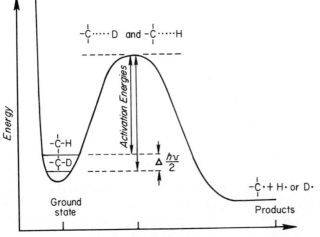

Figure V-5. Analysis of the kinetic isotope effect for C–$_1^1$H and C–$_1^2$H bond-breaking processes. In a reaction in which the rate determining step involves the breakage of a carbon–hydrogen bond, the difference in the rate of the hydrogen and deuterium substituted compounds should correspond to the difference in zero-point energies of these bonds in the ground state.

mum. The difference in activation energy required for the two processes will be equal to the difference in zero-point energy levels; the *lower* C–$_1^2$H zero-point energy level resulting in an energy of activation that is about 1 kcal/mole *greater* than the activation energy for breaking the C–$_1^1$H bond. This difference in the energy of activation results in the normal deuterium kinetic isotope effect where the hydrogen-substituted compound reacts several times faster than the deuterium-substituted compound ($k_H > k_D$). This analysis is summarized in Figure V-5.

This analysis of the kinetic isotope effect shows how the observed rate difference is related to the *relative difference* in mass of the isotopes involved. Deuterium isotope effects (k_H/k_D) are often found to be from 6 to 10, and tritium isotope effects (k_H/k_T) will theoretically be several times larger; but the kinetic isotopic effects involving other atoms of biological interest (carbon, oxygen, nitrogen, sulfur) will be significantly less.*

Since enzymatically catalyzed processes are chemical reactions, biochemical processes may also display $k_H/k_D > 1$, or other reaction rate differ-

* In a recent investigation of a carbon isotope effect, for example, O'Leary found that the enzymatic decarboxylation of glutamic acid displayed a $k^{12}C/k^{13}C$ of 1.0172.[8]

ences due to isotope effects.* For this reason, the isotopic substitution used to *detect* enzymatic biological stereospecificity may also *produce* an observable kinetic isotope effect on the rate of the enzymatic reaction.

Consider, for example, the biochemical reaction of a methylene group ($-CH_2$) at a *pro*-chiral center. Depending upon the enzyme involved, a stereospecifically labeled methylene

will react with biological specificity to lose either the hydrogen or the deuterium atom. If the enzymatic reaction stereospecifically removed a deuterium in the rate-determining step, then this reaction would be significantly slower than the enzymatically catalyzed reaction of the alternatively labeled methylene group,

where the carbon-hydrogen bond would be broken and the hydrogen atom stereospecifically lost ($k_H/k_D > 1$). Note that the isotopic substitution does not alter the absolute stereospecificity of the process, but the isotopic substitution may impose a kinetic isotope effect on the phenomenon of biological differentiation between chemically like, paired groups.†

DIFFERENTIATION OF ENANTIOTOPIC GROUPS BY A CHIRAL CHEMICAL REAGENT

It is not necessary to ascribe biological stereospecificity toward chemically like, enantiotopic paired groups to any particular substrate-enzyme

---

* Additional information on the kinetic isotope effect and its manifestations in reactions of biological interest can be found in W. P. Jencks, *Catalysis in Chemistry and Enzymology*, Chapter 4, McGraw-Hill, 1969.

† See Chapter VI, pages 202–204, for the description of an unusually complex determination of biological stereospecificity involving isotopically labeled substrates in which analysis of the associated kinetic isotope effect is a vital feature of the overall experimental conclusion.

combination, such as three-point attachment. Our analysis of the principles governing this type of stereospecificity indicates that the critical factor will be creation of diastereomeric relationships by interactions of the enantiotopic paired groups with a dissymmetric environment. It follows that an appropriate *chiral chemical reagent* should be able to differentiate between the enantiotopic paired groups at a *meso*-substituted carbon. Schwartz and Carter[5] found that when the *meso*-carbon-containing compound, β-phenylglutaric anhydride, was reacted in homogeneous solution (hot benzene) with the chiral reagent (−)-α-phenylethylamine to form the monoamide, a chemically stereoselective process occurred. Analysis of the monoamide products indicated that two diastereomeric products were produced in a 60:40 ratio.

This chemically stereoselective process (Figure V-6) is analogous to the biologically stereoselective combination of a dissymmetric enzyme with a single enantiotopic branch of a *meso*-carbon center to specifically form one diastereomeric enzyme-substrate complex. Additional chemically stereoselective reactions of *meso*-carbon groupings have been reported. To be sure, the more stringent three-dimensional orientations required for substrate-enzyme combinations in biochemical processes lead to a greatly enhanced stereospecificity. Nevertheless, this observation of a measure of stereoselectivity in a chemical reaction in homogeneous solution indicates that neither a dissymmetric enzyme nor a literal three-point combination is absolutely required to differentiate between chemically like, enantiotopic paired groups. The only critical element is production of diastereomeric relationships by combinations or interactions of the substrate with a dissymmetric agent.

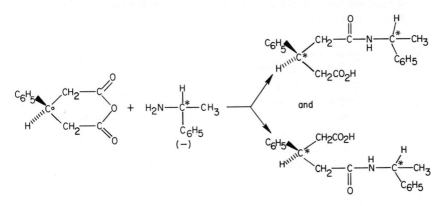

Figure V-6.   Chemical stereoselective reaction of a *pro*-chiral *meso*-carbon-containing compound. (Circle indicates the original *pro*-chiral center; asterisks, the chiral centers).

DIFFERENTIATION BETWEEN DIASTEREOTOPIC PAIRED GROUPS

Thus far in this chapter we have confined our discussion of differentiation between chemically like, paired groups to those groups that occur as paired enantiotopic substituents at a *meso*-carbon type of *pro*-chiral center. Inspection of the structures projected in Figure V-2, however, reveals that only the C-1 carbon of ethanol, the C-3 carbon of citric acid, the C-2 carbon of glycerol, and the methylene carbons of succinate are *meso*-carbon atoms according to the stipulations of Schwartz and Carter. The principles and considerations already developed indicate that biological stereospecific reactions will differentiate between the paired chemical groups at *all types of pro-chiral centers of the representative substrates of Figure V-2*. The methylene hydrogen atoms at the *pro*-chiral C-3 atom of (R)-$(+)$-glyceraldehyde, for example, may be termed nonequivalent, diastereotopic paired hydrogens to distinguish them from the nonequivalent, enantiotopic paired groups previously described. The dissymmetric (R)-$(+)$-glyceraldehyde molecule lacks reflective symmetry. Hence there is no possible reflective symmetry operation that will interchange the two paired hydrogens at this *pro*-chiral center. These hydrogens, therefore, are not enantiotopic. Furthermore, since the (R)-$(+)$-glyceraldehyde molecule also lacks rotational symmetry, there is no possible rotational symmetry operation that will interchange these two paired hydrogens. As is apparent from the total configurational designations, replacement of one of the paired hydrogens at C-3 by a tritium atom would produce $(2R,3S)$-3-tritioglyceraldehyde and replacement of the other would produce the diastereomeric $(2R,3R)$ molecule. Thus both symmetry relationships and replacement tests illustrate that the hydrogens at C-3 of (R)-$(+)$-glyceraldehyde are chemically nonequivalent, diastereotopic paired groups (cf. Table II-3). Diastereotopically related paired groups possess differing orientations relative to the functional groups at the chiral centers, and these paired hydrogen atoms are both chemically and geometrically distinct. To be sure, just as in the case of the diastereotopic paired hydrogens at C-4′ of the dihydronicotinamide rings of NADH and NADPH, these methylene hydrogens of glyceraldehyde are chemically very similar. Still they can potentially be differentiated by any agent that is capable of recognizing the subtle but real chemical differences (220-MHz nmr, e.g.) and stereoselective reactions with one of these paired hydrogens *will not absolutely require a dissymmetric enzyme or a chiral chemical reagent*. Nevertheless, the stringent substrate-enzyme interaction geometries that make *biochemical enantiotopic stereospecificity possible will also make biochemical diastereotopic stereospecificity possible*. Biological processes will, therefore, display a high degree of stereospecificity toward the diastereotopically related paired substituents at such *pro*-chiral centers.

The paired methylene hydrogens at C-2 and C-4 of citric acid (and the analogous methylene hydrogens at C-1 and C-3 of glycerol) provide an instructive example of yet another type of *pro*-chiral center. Inspection of the citric acid molecule in Figure V-2 reveals that the four methylene hydrogen atoms represent paired substituents possessing both enantiotopic and diastereotopic relationships. As previously described, the mirror plane passing through citric acid can be used to illustrate that the C-1 and C-5 carboxyl functions and the C-2 and C-4 methylene carbon atoms possess nonequivalent enantiotopic relationships. This mirror plane also illustrates that the *pro-R* methylene hydrogen of the *pro-S* carboxymethyl branch is enantiotopic to the *pro-S* methylene hydrogen of the *pro-R* branch. The same enantiotopic relationship exists between the *pro-S* hydrogen of the *pro-S* branch and the *pro-R* hydrogen of the *pro-R* branch. The four methylene hydrogens of citric acid thus form two enantiotopically related pairs of hydrogen atoms.

Differentiation between these enantiotopically related substituents will therefore require a dissymmetric reagent and will be the rule in enzyme-catalyzed biological reactions. At the same time, a consideration of the implications of the *pro*-chiral designation reveals that the *pro-R* methylene hydrogen of the *pro-S* branch is diastereotopically related to both the *pro-S* hydrogen of the same *pro-S* branch and to the *pro-R* hydrogen of the enantiotopic *pro-R* branch. A diastereotopic relationship also exists between H_S of the *pro-S* branch, H_R of *pro-S* branch, and H_S of the *pro-R* branch. These diastereotopic hydrogen atoms are chemically and geometrically distinct, and biological stereospecificity between the diastereotopic hydrogens will also be observed in enzymatic reactions.

In Chapter I it was emphasized that the Ogston effect is due in part to the stereospecific hydroxylation of one specific carboxymethyl branch of citrate by the aconitase enzyme. This biological stereospecificity is necessary to explain the specific carboxyl labeling pattern originally noted by Evans and Slotin[9] and Wood et al.[10] (cf. Figure I-3). The foregoing stereochemical analysis demonstrates the nonequivalence of each of the four methylene hydrogens of citrate and indicates that the aconitase reaction should *also* specifically replace a *single unique hydrogen* of the reacting enantiotopic branch.

England utilized the biochemical transformations outlined in Figure V-7 to demonstrate that one of the four chemically like but stereochemically unique methylene hydrogens of citrate is removed in the aconitase reaction.[11] The L-malate product produced by the fumarase (L-Malate hydro-lyase) catalyzed addition of D_2O to fumaric acid must be either the *erythro* (from *anti*-D_2O addition) or the *threo* (from *syn*-D_2O addition) diastereomeric isomers (Figure V-8). Since the two potential products of the fumarase-catalyzed addition of D_2O are diastereomeric, the stereochemistry of the product could be established by the physicochemical properties. Gawron

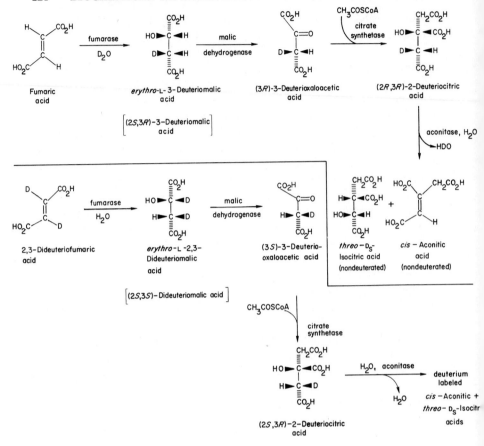

Figure V-7. Demonstration of the stereochemical course of the biological formation of *cis*-aconitate and isocitrate.[11]

and Fondy[12] and Anet[13] independently compared the nmr spectrum of the enzymatic 3-deuteriomalate product with the nmr spectra of stereospecifically labeled 3-deuteriomalates prepared by chemical syntheses. It was concluded in both studies that the enzymatic product was the *erythro*-L-3-deuteriomalate [(2S,3R)-3-deuteriomalate] projected in Figure V-8.* The

---

* Although in theory it is always possible to distinguish diastereomers on the basis of their differing physicochemical properties, the successful identification of a correct stereoisomeric form requires a careful analysis of the experimental data. In this case, for example, the results of an earlier nmr study had been interpreted to indicate that the L-malate product of the fumarase-catalyzed addition of D_2O to fumarate possessed the *threo* relative orientation of groups at the adjacent chiral centers.[14]

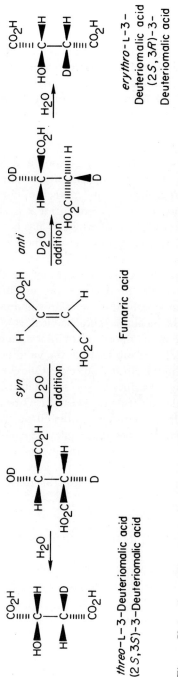
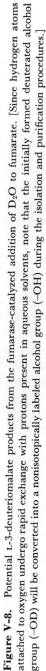

Figure V-8. Potential L-3-deuteriomalate products from the fumarase-catalyzed addition of D_2O to fumarate. [Since hydrogen atoms attached to oxygen undergo rapid exchange with protons present in aqueous solvents, note that the initially formed deuterated alcohol group (—OD) will be converted into a nonisotopically labeled alcohol group (—OH) during the isolation and purification procedures.]

stereospecificity of the H_2O addition to fumarate catalyzed by fumarase was thus determined to be *anti*. As outlined in Figure V-7, England showed that the metabolism of this biosynthetic 3-deuteriomalate [(2S,3R)-3-deuterio-malate] via TCA cycle reactions to *cis*-aconitate resulted in the loss of the deuterium label.[11] In contrast, when the *erythro*-L-2,3-dideuteriomalate* [(2S,3S)-2,3-dideuteriomalate] derived from the fumarase catalyzed *anti* addition of H_2O to 2,3-dideuteriofumarate was subjected to the same sequence of enzymatic reactions, the *cis*-aconitate and isocitrate produced retained deuterium labeling (Figure V-7).[11]

Other experiments carried out by Hanson and Rose† established that the citrate synthetase enzyme stereospecifically utilizes the oxaloacetate substrate to form the *pro-R* carboxymethyl branch of citrate and, furthermore, that the absolute configuration of C-3 of citrate is as shown in Figure V-7.[15] Finally, the configuration of natural isocitrate has been determined to be as indicated in Figure V-7.[16,17] These total results, therefore, show that aconitase stereospecifically replaces H_R of the *pro-R* carboxymethyl branch of citrate to form isocitrate. As predicted from analysis of the chemically like, paired groups present in citrate, the biochemical process differentiates among the four stereochemically unique methylene hydrogen atoms of this substrate.

The combined chemical and biochemical studies by Kun and co-workers[18] on the nature of the fluorocitrate inhibition of aconitase (discussed in detail in Chapter IV) also substantiate the validity of the derived stereochemical relationships in the citrate molecule. On reexamining Figure IV-10 and the accompanying description, we see that the various monofluoro-citrate isomers clearly demonstrate the existence of the predicted enantiotopic and diastereotopic hydrogen positions in the citrate molecule. Furthermore, Kun's observation that the aconitase enzyme combination with mono-fluorocitrate is dependent on one specific monofluorocitrate stereoisomer‡ demonstrates that the biological process manifests the predicted stereospecificity for one structurally unique position at the C-2 and C-4 *pro*-chiral centers of citrate.

THE CHEMICALLY LIKE, PAIRED GROUPS OF SUCCINIC ACID

The structure of the succinic acid molecule provides an example of yet another type of possible *pro*-chiral stereochemical relationship. Each of the

---

* Note here the deficiency of the older *erythro, threo* stereochemical designations. The absolute configuration at C-3 of *erythro*-L-3-deuteriomalic acid is *R*; the configuration at C-3 of *erythro*-L-2,3-dideuteriomalic acid is *S*.

† These experiments are discussed in detail later in this chapter as examples of stereospecific biological reactions at *pro*-chiral trigonal centers, cf. p. 151.

‡ In this case, the stereospecific combination produces inhibition rather than catalysis.

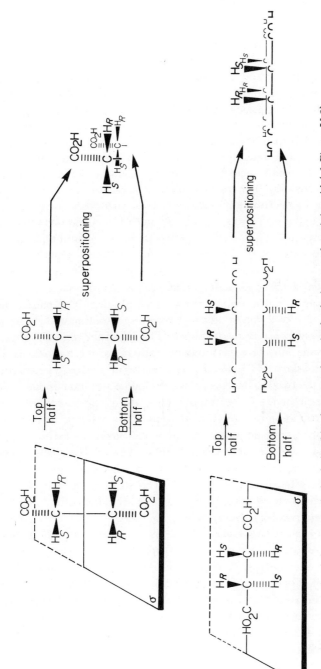

Figure V-9. Analysis of the geometrical relationships between the paired substituents of succinic acid (cf. Figure V-3).

methylene positions of succinic acid may be designated as *meso*-carbon centers (C_{aacd}) according to the Schwartz–Carter definition. However, the presence of symmetry elements in addition to the mirror plane passing through these *meso*-carbon atoms and dividing the paired enantiotopic hydrogens (the mirror plane containing the four carbon atoms) results in a modified set of stereochemical relationships for this molecule. Initially, for example, it might be expected that the presence of the additional mirror plane between C-2 and C-3 perpendicular to the extended carbon chain would, in analogy to the citric acid and glycerol molecules, indicate that the two carboxyl functions of succinic acid bear a nonequivalent enantiotopic relationship. This is not the case. The higher symmetry present in the succinic acid structure results in the superimposability of these two mirror image halves of the molecule (upper and lower)—carboxyl on carboxyl and methylene group on methylene group. Thus, in direct contrast to the enantiotopic relationships within the citric acid molecule, the two carboxyl groups and the two methylene carbon atoms of the succinic acid molecule comprise chemically and geometrically equivalent paired groups. At the same time, a careful examination of the molecule reveals that although the paired methylene groups of the two mirror image halves are superimposable, *the individual hydrogen atoms are not all superimposable*. The H_S of the top half of the succinic acid molecule may be superimposed upon H_S of the bottom mirror image half, and similarly H_R upon H_R; but the two H_S remain geometrically distinct from the two H_R hydrogens. An analogous situation applies with regard to the mirror image halves produced by the plane containing the carbon atoms and dividing the paired hydrogens at the two *meso*-carbons. Again the two mirror halves are superimposable: carboxyl on carboxyl, methylene carbon on methylene carbon, H_R on H_R and H_S on H_S. Again, however, the two H_R hydrogens remain geometrically distinct from the H_S. These relationships are illustrated in Figure V-9. This stereochemical analysis indicates that the paired *carbon atoms* of succinic acid are not enantiotopic but equivalent (superimposable). Nevertheless, the *hydrogen atoms* form a pair of nonequivalent *enantiotopic substituents*. It is predicted, therefore, that although the carboxyl groups of succinic acid are biologically indistinguishable, a dissymmetric enzyme will be capable of differentiating between the pairs of enantiotopic hydrogen atoms. The validity of the first prediction is indicated by the observation that carboxyl label stereospecifically introduced into the TCA cycle from acetate or oxaloacetate becomes randomized at succinate (cf. Figure I-3). The validity of the second prediction is indicated by the stereospecificity manifest in the succinic dehydrogenase reaction (Succinate: oxidoreductase) (cf. pp. 131–133, following).

HIRSCHMANN'S ROTATIONAL SUPERPOSITIONING TEST FOR NONEQUIVALENT PAIRED GROUPS

This stereochemical analysis of succinic acid is correct and leads to valid predictions regarding the potential for biological stereospecificity, but the approach is needlessly complicated. Determination of the geometrical equivalence or nonequivalence of chemically like, paired substituents in substrates is more easily performed by use of the test proposed originally by Hirschmann.[4] Hirschmann's approach emphasizes the possibility of *rotational* superpositioning of paired substituents. Those paired, chemically like, substituents which are not superimposable by rotational symmetry operations (C_n axis, $n > 1$) are nonequivalent and will be subject to biological differentiation. *Those paired substituents which are superimposable by rotational symmetry operations are equivalent and will not be subject to biological differentiation.*

We have seen that chemically like, paired substituents that are *only* interchanged (superimposed) by reflective symmetry operations of a molecule are nonequivalent enantiotopic groups and can be distinguished by chiral agents. Indeed, *since enantiotopic chemical groups must possess the opposite absolute order of arrangement of substituents in space, the demonstration of an enantiotopic relationship between paired substituents is actually a proof of their absolute geometrical nonequivalence.*

Paired groups that are interchangeable by rotational symmetry operations of a molecule, however, must be geometrically and chemically equivalent— even in an absolute sense. Since there is no experimental method—physical, chemical, or biological—whereby equivalent functional groups can be differentiated, such groups will not be subject to biological stereospecific reaction.

The succinic acid molecule will serve as an excellent example of both the ease and the validity of Hirschmann's method of ascertaining the presence or absence of biologically differentiable paired groups. Succinic acid possesses a C_2 axis passing through the C-2 and C-3 carbon–carbon bond, perpendicular to the extended carbon chain. The C_2 rotational symmetry operation interchanges the two carboxyl groups and the two methylene groups, with the final conformation superimposable on the original conformation (by definition for a symmetry operation). Therefore, the two carboxyl groups and two methylene carbons of succinic acid are equivalent and not subject to differentiation by any experimental means. As illustrated in Figure V-10, however, this C_2 rotation *only* interchanges the enantiotopic methylene hydrogens identified in the previous analysis (H_S and H_S; H_R and H_R). Thus Hirschmann's test of rotational superpositioning shows that the two *pro-S* methylene hydrogens are equivalent and that the two *pro-R* hydrogens are

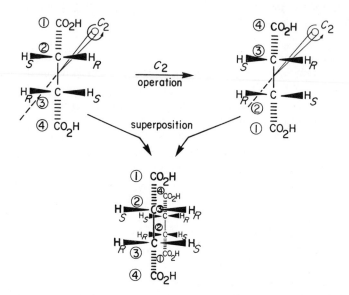

Figure V-10. Rotational superpositioning of equivalent groups in the succinic acid molecule.

equivalent. However, the H_S and H_R pairs, previously identified as enantiotopic, are demonstrated to be geometrically nonequivalent. Biological stereospecificity is observed between the H_S and H_R hydrogens of succinic acid but is impossible between the two equivalent H_S (or H_R) hydrogens.

The validity of this analysis is illustrated by the example in Figure V-11. The biochemist pictured here represents a dissymmetric agent capable of the most subtle and discriminating of stereospecific analyses. He is shown attempting to selectively remove one of the methylene hydrogens of a succinic acid model. This biochemist shares a common property with all dissymmetric enzymes—a specific binding site for the carboxyl group(s) of succinic acid, his left hand; and a specific catalytic site for removing the hydrogen(s), his right hand.

In Figure V-11A, the biochemist has grasped the model with the appropriate binding site (left hand) and is using his catalytic site (right hand) to remove the most readily accessible methylene hydrogen. For illustrative purposes, the carbon atoms of the model have been *artificially numbered.* We see that the agent has arbitrarily grasped the carboxyl group numbered one. In Figure V-11A, the H_S hydrogen on C-2 is properly situated for removal. The arrangement of biochemist and substrate in Figure V-11B represents the situation that prevails after a C_2 rotation of the succinic acid model places the

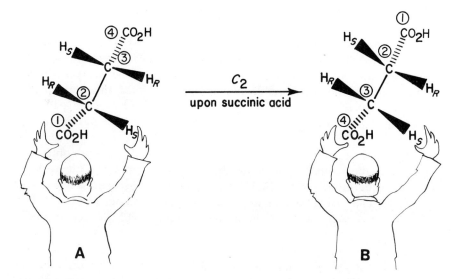

Figure V-11. Attack on a succinic acid molecular model by a biochemist.

carboxyl artificially designated as 4 into the binding site. It can be readily
seen that, except for the artificial carbon chain numbers, the representation
of Figure V-11B is equivalent to that of Figure V-11A. The two illustrations
are totally superimposable: biochemist on biochemist, like chemical group
on like chemical group. Again the hydrogen properly situated for removal by
the catalytically active right hand is an H_S hydrogen (this time the H_S
hydrogen on the carbon arbitrarily numbered three). Thus Figure V-11 shows
how paired groups of a substrate which are superimposable by a rotational
symmetry operation are completely equivalent and cannot be biologically
differentiated. However, those paired groups which are not superimposable
by rotations will be either diastereotopic or enantiotopic. Such chemically
like pairs will be differentiated in biological processes.

   In the first studies of the stereochemistry of the succinic dehydrogenase
reaction, Englard and Colowick determined that both *pro-R* and *pro-S*
hydrogens were removed from succinate during the enzymatic conversion to
fumarate.[19] In view of the geometrical nonequivalence of these hydrogen
atoms, we could assume that the dehydrogenase enzyme was exhibiting the
expected biological stereospecificity and selectively removing H_R and H_S
hydrogens from adjacent carbon atoms. Englard and Colowick's experiment, however, could not distinguish this stereospecific process from the
totally nonselective process in which H_R and H_S hydrogens would both be
eliminated randomly. Tchen and van Milligan reinvestigated the stereo-

chemistry of the succinic acid dehydrogenase reaction and demonstrated that the process is stereospecific rather than random.[20] Racemic and *meso*-2,3-dideuteriosuccinate substrates were prepared by catalytic reduction of fumaric and maleic acids with deuterium gas. Catalytic reductions of the type used are chemically stereospecific reactions involving *syn* hydrogen additions to double bonds. As illustrated in Figure V-12, this stereospecific chemical reduction will produce a racemic 2,3-dideuteriosuccinic acid mixture from fumaric acid and the *meso*-2,3-dideuteriosuccinic acid from maleic acid.

According to the Cahn-Ingold-Prelog system of designating absolute configurations, the racemic 2,3-dideuteriosuccinic acid mixture in Figure V-12 contains equal numbers of (2*R*,3*R*) and (2*S*,3*S*)-2,3-dideuteriosuccinic acid molecules. The *meso*-2,3-dideuteriosuccinic acid can be designated as (2*R*,3*S*)-2,3-dideuteriosuccinic acid. As indicated in Figure V-12, the percentage of dideuterated molecules originally present in the succinic acid diastereomers was determined and compared with the percentage of dideuterated fumarate molecules isolated after incubations with Keilin-Hartree preparations of heart tissue. The *R* and *S* configurational designations emphasize that the *absolute configurations of the adjacent deuterated centers within any single enantiomeric molecule found in the racemic 2,3-dideuteriosuccinic acid mixture are the same.* If the succinic acid dehydrogenase reaction stereospecifically removes H_S and H_R methylene hydrogens from the adjacent carbon atoms of succinate, therefore, *one deuterium must be removed from each of the dideuterated enantiomeric succinate molecules.* As shown, the succinic acid dehydrogenase reaction of a racemic

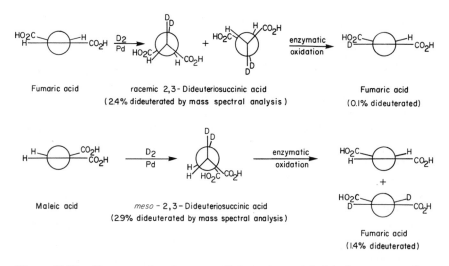

Fumaric acid

racemic 2,3-Dideuteriosuccinic acid
(2.4% dideuterated by mass spectral analysis)

Fumaric acid
(0.1% dideuterated)

Maleic acid

meso - 2,3-Dideuteriosuccinic acid
(2.9% dideuterated by mass spectral analysis)

Fumaric acid
(1.4% dideuterated)

Figure V-12.   Demonstration of stereospecificity in the succinic dehydrogenase reaction.

2,3-dideuteriosuccinate mixture does yield fumarate which lacks dideuterium labeling.

On the other hand, the absolute configurations of the adjacent deuterated centers of the *meso*-2,3-dideuteriosuccinic acid are opposite. If the dehydrogenase reaction removes adjacent enantiotopic H_R and H_S hydrogens from this substrate, it would be anticipated that approximately half of the oxidations would remove the adjacent enantiotopic deuterium atoms and about half would remove the adjacent enantiotopic hydrogen atoms present within *each* dideuterated *meso* molecule.

This observation leads to the prediction that, after this type of stereospecific process, the percentage of dideuterated fumarate molecules should be approximately half the percentage of dideuterated succinate molecules in the original *meso* substrate. Tchen and van Milligan's findings were again consistent with this prediction. Since the experimental results in Figure V-12 are not consistent with a random, nonselective removal of H_R and H_S hydrogens from succinate (all the dideuterated substrates would then have yielded comparable amounts of dideuterated products), the study of Tchen and van Milligan proves that succinic dehydrogenase differentiates between the chemically like, nonequivalent, enantiotopic methylene hydrogens of succinate. * Clearly, these results demonstrate that succinate dehydrogenase is an enzyme which manifests a high degree of stereospecificity towards the succinate substrate molecule. The enzyme, however, exhibits only *the aspects of biological stereospecificity that are permitted by the molecular symmetry of the succinate molecule.* Thus, succinic acid dehydrogenase fails to differentiate between the equivalent paired carboxymethyl functions or the equivalent paired H_S methylene hydrogens. It is unfair to malign succinic acid dehydrogenase and ascribe only two binding sites to this enzyme merely because the enzyme cannot accomplish the impossible (cf. p. 15).

Although Hirschmann's test of the rotational superpositioning of like groups permits a ready determination of the biologically differentiable groups in succinic acid, the application of this approach to such molecules as citric acid is even simpler. In the case of such nondissymmetric molecules as citric acid and glycerol, the *absence of elements of rotational symmetry leads automatically to the conclusion that all the chemically like, paired substituents of these molecules are biologically differentiable.* This conclusion, of course, is the same as that previously derived using the equally valid but more complicated analysis based on reflective relationships and /or the application of *pro*-chiral configurational nomenclature.

---

* The reader is encouraged to determine how the experimental results reported by Tchen and van Milligan may also be used to prove that the reaction catalyzed by succinic dehydrogenase is a stereospecific *anti* elimination process.

THE C$_{aa(+b)(-b)}$ CLASS OF *PRO*-CHIRAL CENTER

Yet another class of saturated *pro*-chiral center is possible. As illustrated in Figure V-13, if *meso*-tartaric acid were to be converted into (2R,4S)-2,4-dihydroxyglutaric acid, the inserted methylene group would comprise a *pro*-chiral center similar to those just described. Upon first examination it might appear that the methylene group of the (2R,4S)-2,4-dihydroxyglutaric acid constitutes a *meso*-carbon center as defined by Schwartz and Carter. The *meso*-carbon, C$_{aacd}$, of Schwartz and Carter, however, has paired chemical groupings (a) and (a), which are inherently nondissymmetric and possess an enantiotopic relationship only because they lie on opposite sides of a molecular mirror plane. In contrast, *pro*-chiral centers of the type just described may be designated C$_{aa(+b)(-b)}$. This designation emphasizes that two types of paired ligands are present. The paired ligands (a,a) are comparable to those associated with a *meso*-carbon center, chemically like and inherently nondissymmetric. The chemically like, paired methylene hydrogen atoms of the 2,4-dihydroxyglutaric acid are such paired (a,a) groups. The (+b) and (−b) paired ligands, however, represent chemically like substituents that are inherently dissymmetric and are, moreover, enantiomeric. The R and S (CO$_2$H)CH(OH)-groupings of the above-mentioned 2,4-dihydroxyglutaric acid correspond to such (+b), (−b) groups. In direct contrast to the *meso*-carbon atom of Schwartz and Carter, the molecular mirror plane of a C$_{aa(+b)(-b)}$ grouping *contains C and the two chemically like (a) groups.* The inherently enantiotopic (+b) and (−b) groups then lie on opposite sides of the mirror plane.

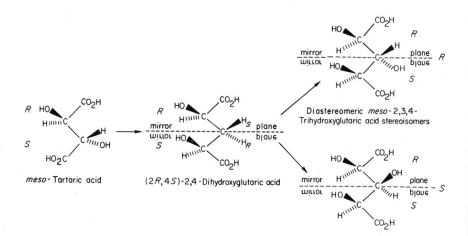

Figure V-13. Conversion of *meso*-tartaric acid into (2R,4S)-2,4-dihydroxyglutaric acid —an example of a C$_{aa(+b)(-b)}$ *pro*-chiral center.

Figure V-13 emphasizes schematically the close relationship that exists between the $C_{aa(+b)(-b)}$ class of *pro*-chiral centers and *meso* molecules. The reader should convince himself that insertion of a methylene group into either of the optically active tartaric acid stereoisomers to form the alternative 2,4-dihydroxyglutaric acid stereoisomers *does not* create any type of *pro*-chiral center. *

According to the arguments and principles developed earlier, the paired *R* and *S* enantiotopic branches of (2*R*,4*S*)-2,4-dihydroxyglutaric acid can be differentiated only by a dissymmetric reagent. A consideration of the stereochemical relationships between the products that would be derived from specific substitutions of each of the methylene hydrogens at the *pro*-chiral center (illustrated by hydroxyl group substitutions in Figure V-13) indicates that these paired hydrogens are diastereotopic. In theory, therefore, the methylene hydrogen atoms of (2*R*,4*S*)-2,4-dihydroxyglutaric acid—or (a) and (a) of any $C_{aa(+b)(-b)}$ *pro*-chiral center—may be differentiated by a chemically selective reagent.

In the following chapter (pp. 189–193) the metabolism of *meso*-α,ϵ-diaminopimelate is described as an example of biological stereospecificity toward the paired groups of such a $C_{aa(+b)(-b)}$ *pro*-chiral center.

It should be noted that specific replacement of one of the paired ligands (a) of a $C_{aa(+b)(-b)}$ *pro*-chiral center does not produce a dissymmetric molecule. The 2,3,4-trihydroxyglutaric acid molecules of Figure V-13, for example, are *meso* molecules possessing reflective symmetry (a mirror plane). Hanson has proposed, therefore, that the *pro*-chiral (a,a) groups of such $C_{aa(+b)(-b)}$ centers be designated *pro-r* and *pro-s* to emphasize the contrast with *pro-R* and *pro-S* groups.[3] He also proposes that the chiral center resulting from replacement of such *pro-r* or *pro-s* ligands be designated as *r* or *s*. Such a system of nomenclature complicates the *pro*-chiral–chiral nomenclature, although adding little to the final stereochemical analysis.

For example, designating the chirality of the trihydroxyglutaric acid in Figure V-13 as (2*R*,3*r*,2*S*) would be dependent on recognition that the compound is a *meso* structure with a mirror plane passing through C-3. It seems reasonable, therefore, to first designate the chiral center at C-3 of these trihydroxyglutaric acids with the *R* or *S* notation by applying the usual sequence rules (which specify that *R* chiral groups have precedence over *S* chiral groups) and *then* to determine that the compounds are optically inactive *meso* molecules. For this reason we have designated all the *pro*-chiral and chiral centers in Figure V-13 with the *R* and *S* nomenclature, avoiding special differentiating designations for the $C_{aa(+b)(-b)}$ class of *pro*-chiral groups.

* In either of these cases, although the 2,4-dihydroxyglutaric acid would be dissymmetric, the paired ligands of the resulting $C_{aa(+b)(+b)}$ grouping are equivalent by the rotational superpositioning test. It follows, therefore, that $C_{aa(+b)(+b)}$ is not a *pro*-chiral grouping.

ASCERTAINING THE PRESENCE OF BIOLOGICALLY
DIFFERENTIABLE PAIRED GROUPS

The following sequence is suggested as an efficient and effective procedure to ascertain the presence of chemically like, biologically differentiable, paired groups within a substrate molecule. First, of course, it should be established that chemically like, paired substituents are present. Since enzymes, as discussed briefly in Chapter I, display a very high specificity for chemically different substituents, we have automatically accepted the principle than an enzyme will readily differentiate a methine hydrogen $\left(\begin{array}{c} \diagup \diagdown \\ -C- \\ | \\ H \end{array}\right)$ from a methylene hydrogen $\left(\begin{array}{c} H \\ | \\ -C- \\ | \\ H \end{array}\right)$ and from a methyl hydrogen $\left(\begin{array}{c} H \\ | \\ -C-H \\ | \\ H \end{array}\right)$, or a primary alcohol group ($-CH_2OH$) from a secondary alcohol group $\left(\begin{array}{c} | \\ -CHOH \end{array}\right)$ and from a tertiary alcohol group $\left(\begin{array}{c} | \\ -C-OH \\ | \end{array}\right)$.

This analysis is concerned with determining whether biological stereospecificity between the pair of methylene hydrogen atoms, or between two equivalently situated primary alcohol groups, is possible. After it has been established that chemically like, paired substituents are present within a substrate molecule, we must project the molecule in the "most symmetrical" accessible conformation and then determine the various reflective and rotational symmetry elements that are present. The absence of all elements of reflective symmetry indicates that the molecule will exist in optically active, enantiomeric forms. Biological stereospecificity will result in the metabolism of one enantiomer or in the formation of one enantiomer from an appropriate *pro*-chiral precursor. The absence of all elements of rotational symmetry (except the trivial C_1 axes) will indicate the nonequivalence of all chemically like, paired groups. Biological stereospecificity will result in the selective reaction of one of these paired substituents. In the event that a molecule does possess elements of rotational symmetry, then Hirschmann's test for the rotational superpositioning of like chemical groups must be applied. Like groups which are superimposed upon one another after a rotational symmetry operation are geometrically and chemically equivalent and cannot be differentiated except by artificial and arbitrary means (e.g., arbitrary numbering systems). Those like groups which are not superimposed by any rotational symmetry operation are geometrically and chemically nonequivalent. Biological stereospecificity between such nonequivalent groups will be observed.

It should be particularly noted that when Hirschmann's test is applied to molecules with rotational symmetry axes above C_2 (C_n, $n \geq 3$), it is necessary to test for rotational superpositioning of the like groups after each possible rotation until a full 360° rotation about the C_n axis has been accomplished. For example, for a C_6 axis, we would have to test for rotational superpositioning after 60, 120, 180, 240, 300, and 360° rotations about the C_6 axis. Superpositioning of the chemically like substituents after any one of these rotational symmetry operations establishes equivalence.

Similar considerations apply when more than one rotational axis can be placed within a molecule. Hirschmann's test of rotational superpositioning must be applied after all possible rotational symmetry operations have been carried out. The case of the allene molecule furnishes a simple example of this observation. The C_2 operation about the rotational axis that is coincident with the carbon–carbon–carbon bonds interchanges (superimposes) the two hydrogens on each of the two terminal carbon atoms (1 and 2; 3 and 4 in Figure V-14). A C_2 rotation about either of the two C_2 rotational axes passing through the central carbon interchanges the hydrogen atoms between the two terminal carbon atoms (1 and 3; 2 and 4 in Figure V-14). Thus, in the case of allene, two rotational superpositioning tests after separate rotational symmetry operations are required to establish the complete equivalence of the four hydrogen atoms.

Since reflective symmetry and rotational symmetry elements occur in various combinations, different aspects of biological stereospecificity will be

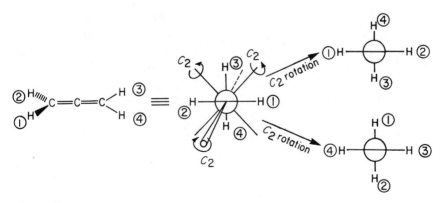

Figure V-14. Application of Hirschmann's test of rotational superpositioning to allene.

manifest with various substrate molecules. For example, a molecule may lack reflective symmetry but possess rotational symmetry. Such a substrate will be dissymmetric and capable of existing in optically active enantiomeric forms. Biological processes will be stereospecific for one of the two possible enantiomers, but within the *single biologically active enantiomer*, rotational superpositioning tests may show that the chemically like, paired groups are equivalent and thus not subject to additional biological differentiation. In contrast, a molecule may possess reflective symmetry but lack all rotational symmetry elements. Such a nondissymmetric molecule does not exist in enantiomeric forms and is not optically active. Obviously, no biological stereospecificity between enantiomers in such a case need be considered. The complete lack of rotational symmetry, however, dictates that the chemically like, paired groups of each substrate molecule will be nonequivalent; thus biological stereospecificity between these paired groups will be observed.

After the presence of biologically differentiable paired groups has been confirmed in a substrate by applying Hirschmann's test of rotational superpositioning, a third step in the stereochemical analysis of a potential substrate is suggested. An analysis of the precise stereochemical relationships that exist between the nonequivalent, chemically like, paired groups is unnecessary to correctly predict the occurrence of biological stereospecificity; however, it will lead to a *more complete understanding* of the overall biological process. Whether an enantiotopic or a diastereotopic relationship exists between the paired groups within a substrate molecule can be established by the appropriate use of chiral and *pro*-chiral designations. Once the configuration of a molecule has been assigned, the stereoisomers resulting from replacement of one of the paired groups with a nondissymmetric test group (e.g., substitution of deuterium for hydrogen) are readily classified as diastereomers or as enantiomers by examining the configurational designations. In this way designating the *pro*-chiral configuration of each paired group that has been established as nonequivalent by Hirschmann's test permits us to ascertain whether differentiation will require a dissymmetric agent (enantiotopic paired groups) or will require only a selective chemical agent (diastereotopic paired groups). Enzymes possess both of these characteristics, and therefore biological processes will typically display extreme stereospecificity between both types of nonequivalent groups. It is important to understand, however, that the principles underlying these two cases of biological differentiation between chemically like, paired groups are only related; they are not identical.

This suggested sequence of steps in the analysis of substrate stereochemistry is illustrated by the examples of threitol and erythritol, which display contrasting stereochemical relationships. The possible erythritol and threitol stereoisomers are depicted in Figure V-15. Inspection of each

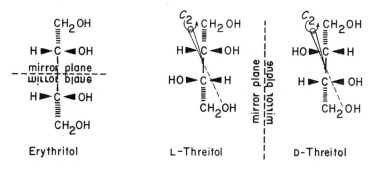

Figure V-15.   Erythritol and threitol stereoisomers.

of these structures immediately establishes that a variety of chemically like, paired functional groups is present within each stereoisomer. These include the terminal hydroxymethyl groups, the secondary alcohol groups, the methylene hydrogens, and the tertiary hydrogens on the carbon atoms bearing the secondary alcohol groups.

The erythritol stereoisomer is related in structure to *meso*-tartaric acid. A mirror plane passes through the molecule between C-2 and C-3 (S_1 axis present). The erythritol stereoisomer, therefore, is not dissymmetric, and enantiomeric molecules will not exist. The threitol stereoisomers, in contrast, lack reflective symmetry and exist in the D and L enantiomeric forms indicated. Biological stereospecificity between these enantiomeric threitol molecules will be observed.

Turning to the biologically differentiable paired groups present within each separate molecule, the lack of a rotational axis of symmetry within the erythritol stereoisomer dictates that all the chemically like, paired substituents will be subject to biological differentiation. In contrast, the two threitol stereoisomers possess C_2 axes passing between C-2 and C-3 perpendicular to the carbon chain. Application of Hirschmann's test of rotational superpositioning to these threitol molecules shows that the two primary alcohol groups, the two secondary alcohol groups, and the two tertiary hydrogens are superimposed by the C_2 rotational operation. These paired chemical functions are therefore equivalent, and biological differentiation between these groups, *within a given threitol enantiomeric substrate*, will be impossible. Hirschmann's test also illustrates that the four methylene hydrogen atoms of the threitol molecules are not all equivalent but can be divided into two differentiable pairs.

The chiral and *pro*-chiral designations of Figure V-16 may now be used to help determine the precise nature of the stereochemical relationships existing between these biologically differentiable paired groups. In the case

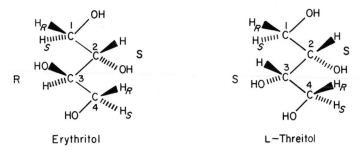

Erythritol   L-Threitol

Figure V-16.   Chiral and *pro*-chiral designations for erythritol and L-threitol.

of the erythritol molecule, the R and S chiral designations indicate that the two secondary alcohol functions (and the similarly situated tertiary hydrogen atoms) are enantiotopic. Since the primary alcohol groups of erythritol lie on opposite sides of the mirror plane, they also possess an enantiotopic relationship. Differentiation between these paired groups will therefore require a dissymmetric reagent or environment. The four stereochemically unique methylene hydrogens of erythritol possess relationships that are analogous to the previously described methylene hydrogens of citric acid. The H_S at C-1 is enantiotopic to H_R at C-4, but diastereotopic to H_R at C-1 and H_S at C-4; whereas H_R at C-1 is enantiotopic to H_S at C-4, but diastereotopic to H_S at C-1 and H_R at C-4. Thus differentiation between the paired methylene hydrogens of any one hydroxymethyl group would theoretically require only a chemically selective reagent.

In the threitol stereoisomer, the rotational superpositioning test indicated that, among all the chemically like groups, only certain of the methylene hydrogens are nonequivalent and subject to differentiation. The chiral and *pro*-chiral designations for the L-threitol molecule (Figure V-16) illustrates that the four methylene hydrogens comprise a pair of *equivalent* H_S hydrogens and a pair of *equivalent* H_R hydrogens. This is similar to the stereochemical relationships among the four methylene hydrogens in the previously analyzed succinic acid molecule. However, in contrast to succinic acid, the H_S and H_R hydrogen pairs in threitol possess a diastereotopic relationship. This is readily ascertained by considering an appropriate substitution test in the L-threitol molecule. Substitution of one methylene hydrogen by tritium would result in a $(1R,2S,3S)$ molecule, whereas substitution of the paired methylene hydrogen would create the diastereomeric $(1S,2S,3S)$ isomer. Differentiation between the nonequivalent H_S and H_R pairs of methylene hydrogens in a threitol molecule will therefore only require a chemically specific reagent. The reader may test his understanding of the stereochemical analyses described here by determining the nature of the relationships present within the D-threitol molecule.

To a limited extent, experiments have established that biochemical processes display the stereospecificities toward erythritol and threitol substrate molecules that are predicted by the foregoing stereochemical analyses. Wawszkiewicz and Barker have found that the metabolism of erythritol by *Propionibacterium pentosaceum* involves the stereospecific phosphorylation of one of the enantiotopic hydroxymethyl groups to form L-α-erythritol phosphate.[21] In contrast, Batt et al. have determined that the two hydroxymethyl groups of L-threitol cannot be differentiated in biochemical reactions.[22] These workers investigated the biological stereospecificity manifest by the xylitol dehydrogenase reaction [xylitol:NAD oxidoreductase (D-xylulose)]. As indicated in Figure V-17, this enzymatic system catalyzes reversible reduction-oxidation conversions of L-erythrulose and L-threitol.

This enzymatic reaction was first used to convert *specifically labeled L-erythrulose-4-*[14]*C* into L-threitol labeled with ^{14}C in a terminal hydroxymethyl grouping. This L-threitol-^{14}C then served as a substrate for the reverse reaction: conversion of L-threitol-^{14}C to L-erythrulose-[14]C. It was experimentally established that the *L-erythrulose-*[14]*C product finally isolated* was about *equally labeled in C-1 and C-4* with [14]C. This experimental sequence, as outlined schematically in Figure V-17, establishes that the L-threitol-^{14}C, formed from L-erythulose-4-^{14}C, is metabolized as L-threitol-1,4-[14]C. As predicted by stereochemical analysis of the threitol structure, no experimental differentiation between the equivalent terminal hydroxymethyl functions of L-threitol is possible—even in an enzymatic system.

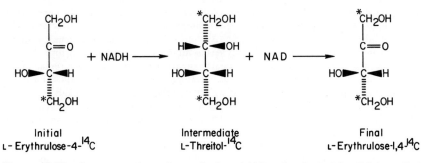

Figure V-17. Interconversions of L-erythrulose-4-^{14}C and L-threitol-1,4-^{14}C by xylitol dehydrogenase (^{14}C-labeled positions are indicated by asterisks).

TRIGONAL *PRO*-CHIRAL GROUPS:
THE *RE/SI* NOMENCLATURE OF HANSON

To this point our discussion of biological stereospecificity has been confined to potential substrates in which the *pro*-chiral centers consist of saturated sp^3-hybridized carbon atoms. However, a large number of biochemical substrates contain *pro*-chiral carbon atoms that are sp^2 hybridized. Aldehydes and ketones are among the most common type of biological substrates that contain such *trigonal pro-chiral carbon centers*. However, the carbonyl functional group of aldehydes and ketones is only one representative of a larger group of trigonal groupings (X_{abc}) which can manifest biological stereospecificity. The trigonal X_{abc} grouping at a carbonyl carbon atom is planar and the aldehyde or ketone may possess a molecular mirror plane coincident with the plane defined by this trigonal carbonyl group. Such molecules will therefore be nondissymmetric and will not be capable of existing in optically active forms.* As illustrated in Figure V-18, however, simple acyclic aldehydes and ketones will lack elements of rotational symmetry unless the (a) and (b) substituents are chemically equivalent [formaldehyde (H_2CO) or a "symmetrical" ketone such as acetone ($CH_3)_2CO$].

The symmetry of the acetaldehyde and oxaloacetic acid molecules in Figure V-18 is the same as the symmetry previously described for the glycerol and citric acid molecules; that is, a σ mirror plane is present but the molecules lack elements of rotational symmetry. The two faces of the trigonal groups (top and bottom of the π bonds) are interchanged (superimposed) by the reflective symmetry operations; but, of course, they cannot be interchanged by the nonexistent rotational symmetry operations. The trigonal carbonyl groups of such simple aldehyde and ketone molecules, therefore, are related in a fundamental way to the *meso* type of saturated *pro*-chiral

---

* It should be noted that we are not concerned here with carbonyl groups present in dissymmetric molecules. Optical activity *is* associated with the carbonyl chromophore of dissymmetric molecules. In the steroids, for example, the optical activity associated with carbonyl absorptions has been extensively used to determine molecular configurations. In such cases, the planar carbonyl group acts as an inherently nondissymmetric chromophore which is perturbed by the dissymmetric environment created by the remainder of the molecule. (See C. Djerassi, *Optical Rotatory Dispersion*, McGraw-Hill, 1960, for a discussion of the optical activity associated with the carbonyl absorption bands of dissymmetric molecules.) In this book, however, we are concerned with trigonal groups present in nondissymmetric molecular structures. The carbonyl groups of acyclic ketones that can readily attain a planar molecular conformation are a common example of such trigonal groups. In these cases, a dissymmetric molecular environment, which perturbs the planar carbonyl group, is absent and the molecule will be optically inactive. Also, in such cases biological stereospecificity toward the opposing sides of the planar trigonal center cannot be ascribed to the influence of the surrounding molecular environment and must be explained in terms of the inherent structural features of the trigonal center itself.

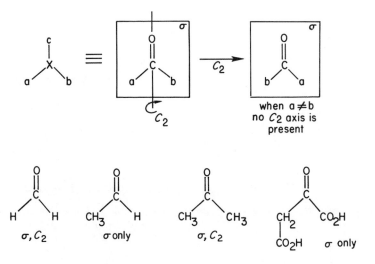

Figure.V-18   Representative molecules potentially possessing trigonal *pro*-chiral centers.

carbon atoms in glycerol and citric acid. Just as the nonequivalent enantio-topic branches of glycerol and citric acid were found to be biologically differentiable, the nonequivalent enantiotopic faces (top and bottom in Figure V-18) of the trigonal carbonyl groups of acetaldehyde and oxalo-acetic acid are subject to biological differentiation.

We can readily establish that such trigonal carbonyl groups are legiti-mate *pro*-chiral centers with enantiotopic faces by considering the substitu-tion tests outlined in Figure V-19. As indicated, a stereospecific reduction with D^- from above the plane of the oriented acetaldehyde molecule pro-duces the (*S*)-1-deuterioethanol, whereas reduction from below the plane produces the enantiomeric (*R*)-1-deuterioethanol. The carbonyl group of acetaldehyde thus satisfies the definition of a *pro*-chiral center, and it is demonstrated by a replacement test that the two faces of the carbonyl group lying on opposite sides of the mirror plane are nonequivalent and enantio-topic. Because of the enantiotopic relationship, differentiation between the two nonequivalent faces of the carbonyl group will require a dissymmetric reagent or environment.

Before biological stereospecificity toward X_{abc} trigonal *pro*-chiral centers can be discussed in a meaningful manner, it is necessary to develop an appro-priate method for designating the two nonequivalent faces. The system of *re/si* stereochemical nomenclature suggested by Hanson for such trigonal atoms[3] is readily applicable and is presented here. This system is based on the Cahn-Ingold-Prelog method of absolute configurational nomenclature

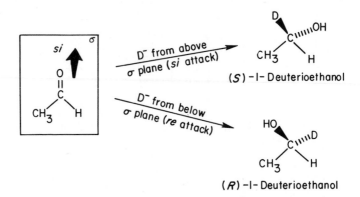

Figure V-19. Stereospecific reactions (reduction) at the *pro*-chiral carbonyl group of acetaldehyde.

and utilizes the priority rules previously described (pp. 70–73). Application of the method consists of three steps:

1. The priorities of the three groups attached to the X_{abc} trigonal atom are determined.

2. The molecule is then oriented so that the observer is either above or below the plane of the trigonal grouping.

3. If the priority sequence of the three attached ligands in order of decreasing priority (a > b > c) determines a clockwise rotation, then the *re* (for *rectus*) face of the trigonal atom is closest to the observer.

If the priority sequence determines a counterclockwise rotation, then the *si* (for *sinister*) face of the trigonal atom is closest to the observer.

We can thus specify whether a biological process involves stereospecific addition to the *re* or to the *si* face of a *pro*-chiral X_{abc} trigonal center. Note, however, that the actual chirality of the product formed by addition to a trigonal center is dependent on the specific functional group added. The *re* and *si* additions of CN^-, instead of D^-, to the acetaldehyde molecule in Figure V-19 would produce cyanohydrins with configurations opposite to those of the illustrated deuterioethanols.

Application of the *re/si* nomenclature to different types of trigonal X_{abc} centers is illustrated in Figures V-20–V-24.

The *re/si* faces of the planar methylethylisopropylborane are readily assigned by use of the rules and steps just outlined [$CH(CH_3)_2$ > C_2H_5 > CH_3]. Since applying the *re/si* nomenclature involves establishing the priorities of the *three substituents actually attached to the trigonal center X*, in contrast to the normal Cahn-Ingold-Prelog treatment of multiple bonds, multi-

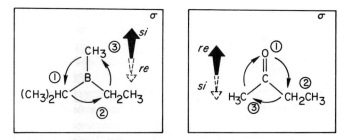

Figure V-20.   Application of *re/si* nomenclature.

ple bonds to the center X itself are *not replicated*. This modification of the Cahn-Ingold-Prelog nomenclature system is illustrated in the case of methyl ethyl ketone ($O > C_2H_5 > CH_3$).

Compounds containing two trigonal centers attached by a double bond will generally possess nonequivalent faces corresponding to the two opposed sides of the plane of the $X_{ab}=Y_{cd}$ assembly. The two faces of such assemblies are readily designated by applying the *re/si* nomenclature separately to the two trigonal centers X and Y. This is illustrated in Figure V-21 for the *cis* and *trans*-1-bromo-1-propene molecules. Although it is unnecessary for the examples of Figure V-21, in some cases the multiple bond must be expanded with replica atoms in order to properly assign the priorities of the three ligands (a, b, Y and c, d, X) attached to each of the trigonal centers.

The opposed faces of a trigonal atom bonded to only two different

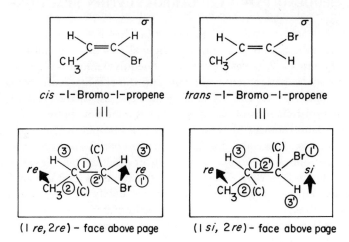

Figure V-21.   Application of *re/si* nomenclature to $X_{ab}=Y_{cd}$ systems.

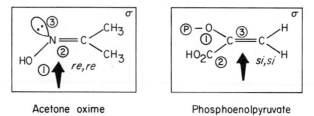

Acetone oxime   Phosphoenolpyruvate

Figure V-22. Application of *re/si* nomenclature to $X_{aa}=Y_{cd}$ assemblies. In this and subsequent illustrations Ⓟ is used to indicate a phosphoryl (PO_3H_2) group.

groups (X_{aac}) are generally equivalent (superimposable by rotation) and hence indistinguishable. However, in the case of assemblies where two trigonal centers are connected by a double bond, such as those illustrated in Figure V-22, the opposed faces of the X_{aac} assembly will be geometrically distinct. In such cases, the $X_{aa(Y)}$ faces of the $X_{aa}=Y_{cd}$ assembly are assigned the *re* or *si* stereochemistry that is determined by the attached $Y_{cd(X)}$ center. The acetone oxime molecule in Figure V-22 illustrates this rule. The acetone oxime molecule also illustrates that a lone pair of electrons can serve as one of the three nonequivalent ligands of a trigonal center if the chemical grouping has the appropriate overall geometry.

BIOLOGICAL STEREOSPECIFICITY TOWARD TRIGONAL (UNSATURATED) *PRO*-CHIRAL CENTERS: STEREOSPECIFICITIES OF PHOSPHOENOLPYRUVATE CARBOXYLATION REACTIONS

Phosphoenolpyruvate is an important biochemical intermediate possessing an $X_{aa}=Y_{cd}$ type of unsaturated, trigonal *pro*-chiral assembly. As illustrated in Figure V-22, the geometrically nonequivalent faces of phosphoenolpyruvate may be designated as the *re,re* and the *si,si* faces, according to the rules formulated previously. By making effective use of the established biological stereospecificities of the glucose phosphate isomerase (D-Glucose-6-phosphate ketol-isomerase) and enolase (2-Phospho-D-glycerate hydrolyase) reactions, Rose and his colleagues prepared the stereospecifically labeled 3-tritiophosphoenolpyruvate substrates pictured in Figure V-23 (cf. Chapter VI, pp. 250–254). These substrates were then used to establish that the carboxylation of phosphoenolpyruvate catalyzed by phosphoenolpyruvate carboxylase [Orthophosphate:oxaloacetate carboxy-lyase (phosphorylating)] from peanuts and *Acetobacter xylinum*, phosphoenolpyruvate carboxykinase [GTP:oxaloacetate carboxy-lyase (transphosphorylating)] from pigeon liver, and by phosphoenolpyruvate carboxytransphosphorylase

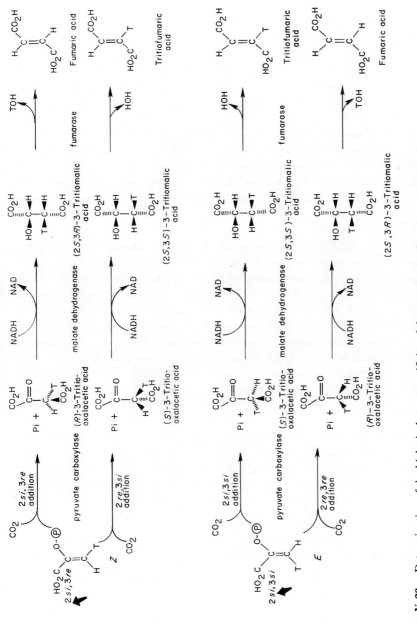

Figure V-23. Determination of the biological stereospecificity of the pyruvate carboxylase reaction.

147

[Pyrophosphate:oxaloacetate carboxy-lyase (phosphorylating)] from *Propionibacterium shermanii, all* involved the stereospecific addition of carboxyl to the *si,si* face of the *pro*-chiral methylene carbon of phosphoenolpyruvate.[23] As illustrated in Figure V-23 for the phosphoenolpyruvate carboxylase reaction, the stereospecific carboxylations of each of labeled 3-tritiophosphoenolpyruvate substrates will lead to stereospecifically labeled 3-tritioxaloacetate products. Each of the enzymatic carboxylation reactions was carried out in the presence of malate dehydrogenase and NADH. In this way, the initial 3-tritioxaloacetate products were rapidly converted to L-malate before enolization and racemization at the labeled methylene carbon (C-3) of the initial oxaloacetate product could occur. The 3-tritiomalate product of each of the coupled enzymatic carboxylations and reductions was isolated and converted to fumarate by a final incubation with fumarase. Since it was known that fumarase catalyzes the stereospecific loss of the *pro-R* methylene hydrogen from L-malate (cf. pp. 123–126), the amount of tritium label lost into the medium during the fumarase incubation could be used to establish the amount of L-(3R)-3-tritiomalate [(2S,3R)-3-tritiomalate] in the original isolated samples. In the case of the products of the phosphoenolpyruvate carboxylase (peanut) incubation, when the original tritium label was *cis* to the phosphate group,* 92.5% of the tritium label was lost from the 3-tritiomalate during the enzymatic conversion to fumarate. As illustrated in Figure V-23, this result indicated that the 3-tritioxaloacetate intermediate was (R)-3-tritioxaloacetate and was derived from an initial stereospecific carboxylation on the (2si,3re) face of the 3-tritiophosphoenolpyruvate substrate (corresponding to the *si,si* face of unlabeled phosphoenolpyruvate). When, in contrast, the initial tritium label in the 3-tritiophosphoenolpyruvate was located *trans* to the phosphate group (E isomer), only 7% of the tritium label was lost from the 3-tritiomalate during the enzymatic conversion to fumarate. Figure V-23 shows that this result indicated an intermediate (S)-3-tritioxaloacetate which was derived from an initial stereospecific carboxylation on the (2si,3si) face of the 3-tritiophosphoenolpyruvate substrate (again corresponding to the *si,si* face of unlabeled phosphoenolpyruvate). Similar experimental results were obtained with the 3-tritiomalate samples derived from the other carboxylation reactions investigated.†

* The configuration at the double bond in molecules such as the tritiated phosphoenolpyruvates of Figure V-23 is readily designated by another application of the Cahn-Ingold-Prelog priority rules. If the two groups of highest priority at the ends of the double bond are on the same side of the double bond (*cis*), the double bond configuration is designated by Z (for *Zusammen*). When the two groups of highest priority are on opposite sides of the double bond (*trans*), the double bond configuration is designated by E (for *Entgegen*).

† In the case of the phosphoenolpyruvate carboxytransphosphorylase from *P. shermanii*, the experimental results indicated some lack of overall stereospecificity. This observation lead Rose et al. to suggest that this enzymatic carboxylation reaction might involve a

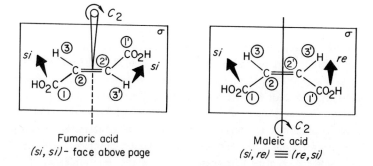

Fumaric acid	Maleic acid
(si, si)- face above page	*(si, re)* ≡ *(re, si)*

Figure V-24. Fumaric and maleic acids as contrasting examples of X_{ab}=X_{ab} trigonal assemblies.

Rose et al. were thus able to establish that each of these enzymatic carboxylations occurred by a stereospecific addition of carboxyl to C-3 of phosphoenolpyruvate from the *si,si* face of the trigonal *pro*-chiral assembly.

BIOLOGICAL STEREOSPECIFICITY TOWARD TRIGONAL (UNSATURATED) *PRO*-CHIRAL CENTERS: STEREOSPECIFICITY OF THE FUMARASE REACTION

The fumaric and maleic acids in Figure V-24 furnish additional instructive and contrasting examples of the stereochemical possibilities associated with trigonal assemblies. Both molecules contain symmetry elements in addition to the σ plane that is coincident with the plane of the X_{ab}=X_{ab} trigonal assembly. In particular, both molecules contain a C_2 rotational axis which permits the rotational superpositioning of each of the chemically like, paired ligands. Thus, for example, the two trigonal carbon atoms (X, X) within each molecule are equivalent and cannot be experimentally differentiated. As illustrated in Figure V-24, the C_2 axis of fumaric acid is *perpendicular* to the plane of the X_{ab}=X_{ab} assembly. The C_2 operation, therefore, does not interchange the *two faces* of the trigonal assembly which lie on opposite sides of this plane. Hence the *si,si* face of fumaric acid is geometrically distinct from the *re,re* face, even though the two trigonal atoms themselves (and the remaining paired groups) are totally equivalent (cf. Figure V-25). In con-

---

transient form of bound pyruvate (cf. Ref. 23, p. 6131). Nevertheless, the results ($\cong 86\%$ tritium lost from one malate sample versus $\cong 30\%$ from the second malate sample) obtained from the experiments with phosphoenolpyruvate carboxytransphosphorylase did clearly indicate a preponderance of stereospecific *si,si* carboxylation.

trast, the C_2 rotational axis of the maleic acid molecule *lies in the plane* of the X_{ab}=X_{ab} assembly. Therefore the same C_2 operation that interchanges the two trigonal carbon atoms, the two vinyl hydrogens, and the two carboxyl groups of maleic acid will also cause the *rotational superpositioning of the two faces* of the trigonal assembly. The *re,si* and *si,re* faces of maleic acid are thus equivalent and cannot be experimentally differentiated.

Enzymatic reactions of the TCA cycle furnish several instances of biological stereospecificity toward trigonal *pro*-chiral centers. For example, the experiments of Gawron and Fondy[12] and Anet[13] described previously (pp. 123–124) established that fumarase catalyzes the *anti* addition of water to fumarate to yield L-malate. Even though the paired *chemical groups* of fumarate are superimposable by rotation and therefore equivalent, this experimental result establishes that the fumarase reaction is nevertheless stereospecific. As shown in Figure V-25, this result requires that fumarase catalyze the addition of hydroxyl to the geometrically unique *si,si* face and the addition of hydrogen to the unique *re,re* face of fumarate. Thus, even though the carbon atoms of fumarate undergoing addition (C-2 and C-3) are indistinguishable, the hydroxyl added to *either of these equivalent positions approaches from only one enantiotopic face* of the X_{ab}=X_{ab} trigonal assembly.

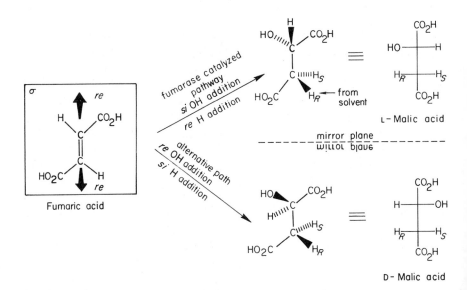

Figure V-25. Stereospecificity of fumarase-catalyzed hydration of fumarate.

BIOLOGICAL STEREOSPECIFICITY TOWARD TRIGONAL (UNSATURATED) *PRO*-CHIRAL CENTERS: STEREOSPECIFICITIES OF CITRATE SYNTHETASE AND ACONITASE REACTIONS

The initial reaction of the TCA cycle also manifests biological stereo-specificity toward a trigonal *pro*-chiral center. This initial reaction involves the condensation of acetyl CoA at the trigonal carbonyl of oxaloacetate. Owing to the nature of the group that becomes attached to the trigonal center (a second carboxymethyl group), the citrate product will contain a central *pro*-chiral center instead of actually being chiral. Nevertheless, the trigonal carbonyl of oxaloacetate is an unsaturated *pro*-chiral center with enantiotopic *re* and *si* faces. It is to be expected, therefore, that citrate synthetase will catalyze the reaction in a stereospecific manner involving either *re* or *si* attack of acetate on the carbonyl group, so that *one* of the carboxymethyl groups attached to the C-3 *pro*-chiral center of the citrate product will be uniquely derived from acetyl CoA. Figure V-26 illustrates the stereochemical possibilities inherent in the biological formation of citrate.

Although the aconitase enzyme stereospecifically hydroxylates one geometrically distinct branch of citrate to produce *threo*-D_s-isocitrate (cf. Figure III-26), the differential labeling pattern observed in α-ketoglutarate following TCA cycle metabolism (cf. Figure I-3) *also requires a stereospecific citrate synthetase reaction*. Figure V-26 illustrates how a *random re* and *si* addition of labeled acetate to the oxaloacetate carbonyl would yield citrate labeled in *both* the carboxymethyl branches. The stereospecific action of aconitase on a citrate substrate randomly labeled in the paired branches could not produce the differentially labeled *threo*-D_s-isocitrate required as the precursor of differentially labeled α-ketoglutarate. Thus the total Ogston effect is dependent both on a stereospecific citrate synthetase reaction and on a stereospecific aconitase reaction. Appropriate experiments utilizing tritium and ^{14}C-labeled substrates readily established that the methylene hydrogens originally present in the oxaloacetate molecule were lost during the biological conversion of the citrate to α-ketoglutarate, whereas those methylene hydrogens derived from acetate were retained (cf. Figure V-7). This shows that the formation of citrate by citrate synthetase was indeed stereospecific and followed one of the stereochemical paths outlined in Figure V-26. Establishing the absolute stereochemistry of the citrate con-densation (*re* or *si* pathway), however, was more difficult. After two indirect methods led to contradictory conclusions, Hanson and Rose established that the *pro-R* carboxymethyl branch of citrate formed by the usual citrate synthetase enzyme was uniquely derived from the oxaloacetate substrate.[15]

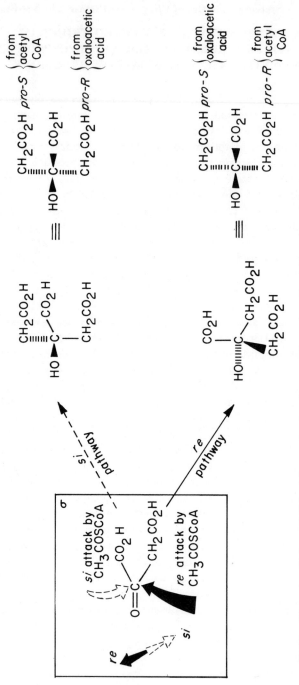

Figure V-26. Stereochemical possibilities inherent in the citrate synthetase-catalyzed formation of citrate.

The biological condensation therefore follows the stereospecific *si* pathway outlined in Figure V-26. To establish this stereochemical course of the condensation, citrate stereospecifically labeled with tritium in the *pro-R* branch [(2R,3R)-2-tritiocitrate] was prepared. When acted upon by aconitase, 95 to 97% of the tritium label in this substrate was lost in the form of tritiated water. Since the portion of the citrate molecule derived from oxaloacetate was known to bear the hydrogen that was exchanged with water by aconitase, this result definitively established the absolute stereochemical course of citric acid biosynthesis.

Obviously the key aspect of Hanson and Rose's determination was the preparation of the (2R,3R)-2-tritiocitrate. This stereospecific synthesis was accomplished with the aid of an extract from *Aerobactor aerogenes*, which contained quinate dehydrogenase (Quinate:NAD oxidoreductase) and 5-dehydroquinate dehydratase (5-Dehydroquinate hydro-lyase) activities. The incubation of either 5-dehydroshikimate or quinate in tritiated water in the presence of the active *A. aerogenes* extract and the appropriate cofactors yielded 6-tritioquinate as outlined in Figure V-27. The enzymatically prepared and tritiated quinate was then chemically oxidized to citrate. Since the absolute configuration of quinic acid was known to be that projected in Figure V-27, the total biochemical and chemical sequence must produce the (3R)-2-tritiocitric acid illustrated.

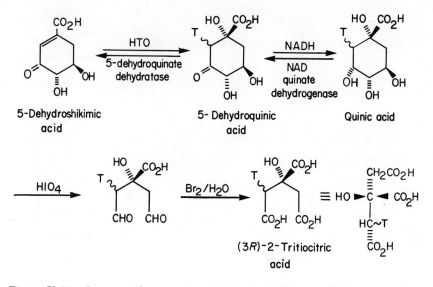

Figure V-27. Stereospecific conversion of 5-dehydroshikimate to (3R)-2-tritiocitrate.[15] (∼T indicates that the absolute configuration of tritium in the products is not yet specified).

Previously, we have described how the experiments performed by England (cf. Figure V-7) related the stereochemistry of the fumarase reaction with that of the aconitase reaction.[11] England demonstrated that the deuterium originally present in L-*erythro*-3-deuteriomalate was stereospecifically removed during the conversion of citrate to isocitrate. As shown in Figure V-7, this partially established the configuration of the hydrogen atom lost in the aconitase reaction. Prior to the experiments of Hanson and Rose, however, it was impossible to ascertain whether the aconitase reaction exchanged the 2R hydrogen of the *pro-R* branch or the 2R hydrogen of the *pro-S* branch of citrate (Figure V-28). The combined experiments of England and of Hanson and Rose established that the deuterium is stereospecifically lost from (2R,3R)-2-deuteriocitrate during the aconitase reaction.

Hanson and Rose also prepared biologically labeled citrate by incubating isocitrate in tritiated water in the presence of aconitase. When this (2R,3R)-2-tritiocitrate was reincubated in H_2O with aconitase, 95% of the tritium was lost in the form of water. Since comparable results were obtained when the labeled citrates derived from biochemically tritiated quinate were incubated with aconitase, these labeled citrates (Figure V-27) must also be assigned the (2R,3R)-2-tritiocitrate configuration. This conclusion, in turn, permits the stereochemical course of the 5-dehydroquinate dehydratase re-

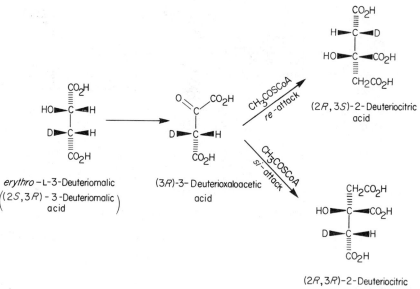

Figure V-28.   Possible citrate molecules from L-*erythro*-3-deuteriomalate.

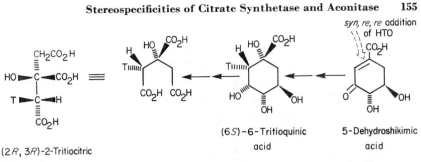

Figure V-29.   Deduction of stereochemical course of the 5-dehydroquinate dehydratase reaction.

action to be determined. As illustrated in Figure V-29, isolation of $(2R,3R)$-2-tritiocitric acid from the tritiated quinate requires that this acid be $(6S)$-6-tritioquinic acid. This establishes that the 5-dehydroquinate dehydratase enzyme stereospecifically catalyzes the *syn* addition of water to the *re,re* face of the carbon—carbon double bond of 5-dehydroshikimate.

Finally, determining the absolute configuration of the hydrogen atom removed from citrate during the conversion to *cis*-aconitate establishes that the enzymatic addition of elements of water to *cis*-aconitate to form citrate is an *anti* process, involving hydroxyl addition on the *re* face of the trigonal C-3 carbon (Figure V-30). The aconitase-catalyzed interconversion of isocitrate and *cis*-aconitate can be assigned as an *anti* addition-elimination process solely on the basis of the established configuration of "natural isocitrate" (cf. Figure III-26). Figure V-30 emphasizes that the aconitase stereospecificity described previously necessitates the addition of both hydrogen and hydroxyl to a given carbon atom of *cis*-aconitate from the *same side of the double bond*. Addition of hydroxyl to the *re* face of C-3 and hydrogen to the *re* face of C-2 yields citrate; the addition of hydrogen to the *re* face of C-3 and hydroxyl to the *re* face of C-2 yields the correct isocitrate isomer (*threo*-D_s-isocitrate). Obviously these stereochemical results are pertinent to the mechanism of the aconitase reaction (cf., e.g., Ref. 24).

It is interesting to note that after Hanson and Rose had established the stereospecificity of the usual citrate synthetase enzyme, a group of anaerobic bacteria was found to possess an atypical citrate synthetase enzyme manifesting the opposite biological stereospecificity.[25–27] Since the usual enzyme carries out a stereospecific *si* addition of acetyl CoA to the trigonal ketone center of oxaloacetate, the two enzymes can be readily described as the *si*-citrate synthetase enzyme and the atypical *re*-citrate synthetase. Because it has been established that (S)-citryl CoA is an intermediate in the biochemical condensation catalyzed by the usual *si*-citrate synthetase, most authors

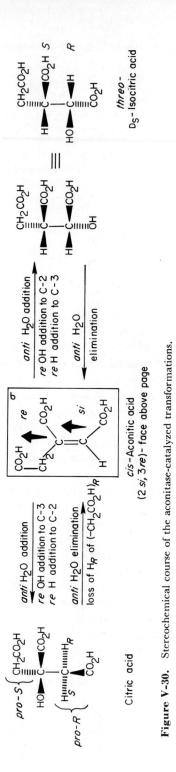

Figure V–30. Stereochemical course of the aconitase-catalyzed transformations.

designate this enzyme as (S)-citrate synthetase and the atypical re-citrate synthetase as the (R)-synthetase. The student should be familiar with both possible designations and, more important, he should recognize that the (R)-citrate synthetase would produce $(3S)$-citrate-2-^{14}C from oxaloacetate-3-^{14}C and acetyl CoA.

The presence of an uncommon citrate synthetase with reversed stereospecificity in some organisms was established by taking advantage of the known stereospecificity of another enzyme. The citrate lyase enzyme (citrate oxaloacetate-lyase) cleaves citrate to acetate and oxaloacetate. It was known that when citrate formed by the normal synthetase enzyme from acetyl CoA and ^{14}C-labeled oxaloacetate was incubated with citrate lyase, the oxaloacetate released contained the ^{14}C label and the cleaved acetate was not radioactive. Based on the results of Hanson and Rose[15] then, we can describe the citrate lyase enzyme as stereospecifically converting the pro-S branch of citrate into the acetate product. The C-3 grouping and the pro-R branch then form the oxaloacetate product. The biochemical sequence in Figure V-31 can therefore be utilized to determine the absolute stereospecificity of the citrate synthetase reactions (the conversion of the original oxaloacetate product to L-malate is carried out only to permit a more convenient analysis of the oxaloacetate label).

Table V-1 presents experimental results indicating that two strains of *Desulfovibrio desulfuricans* and two strains of *D. vulgaris* contain the citrate synthetase with the unusual re [or (R)] stereospecificity. *Clostridium kluyveri*, *C. acidi-urici, and C. cylindrosporum* have also been demonstrated to possess the re-citrate synthetase.[25,26] As Table V-1 illustrates, however, the si-citrate synthetase characteristic of mammalian systems (the enzyme purchased from Boehringer) and aerobic bacteria is also the usual citrate synthetase of a number of the anaerobic or photosynthetic bacteria.

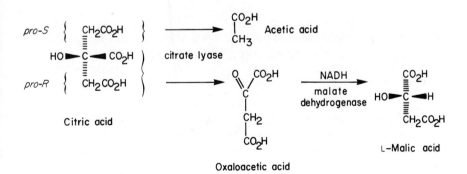

Figure V-31. Determination of the absolute stereochemistry of citrate synthetase reactions via the citrate lyase reaction.

Table V-1   Cleavage of Citrate-^{14}C Samples with Citrate Lyase[27]

Source of citrate [a]	Radioactivity in reaction products	
	% in Acetate	% in Malate
Citrate synthetase (Boehringer)	0.2	99.8
D. desulfuricans (8307)	97.0	3.0
D. desulfuricans (8372)	97.9	2.1
D. gigas (9337)	0.1	99.9
D. salexigens (8403)	0.2	99.8
D. vulgaris (8303)	99.2	0.8
D. vulgaris (8386)	98.9	1.1
Dm. ruminis (8452)	1.8	98.2
Rhodops. capsulata	0.1	99.9
Rhodops. palustris	0.3	99.7
Rhodops. spheroides	0.1	99.9
Rhodops. rubrum	0.6	99.4

[a] The citrate-^{14}C samples were prepared from oxaloacetate-4-^{14}C and acetyl CoA with cell free extracts of the indicated microorganisms. Aliquots of the citrate-^{14}C samples were then stereospecifically degraded as outlined in Figure V-31.

The evolutionary implications of this interesting case of comparative biology has been discussed by Gottschalk.[27] The investigation of *si* and *re*-citrate synthetases has been complicated by reports that the stereospecificities of the citrate synthetases of *Clostridium kluyveri*, *C. acidi-urici*, and *C. cylindrosporum* are altered by the methods used to prepare the active enzyme extracts.[28] (However, cf. Ref. 29).

BIOLOGICAL STEREOSPECIFICITY TOWARD TRIGONAL (UNSATURATED) *PRO*-CHIRAL CENTERS: STEREOSPECIFICITIES OF MALATE DEHYDROGENASE AND ALCOHOL DEHYDROGENASE REACTIONS

Additional common examples of biological stereospecificity toward the geometrically nonequivalent *re* and *si* faces of a trigonal center are the carbonyl reductions catalyzed by various dehydrogenases. Malate dehydrogenase (L-Malate:NAD oxidoreductase), for example, must catalyze the stereospecific addition of hydride from NADH to the *re* face of the oxaloacetate ketone group in order to produce only L-malate [(*S*)-malate]. The

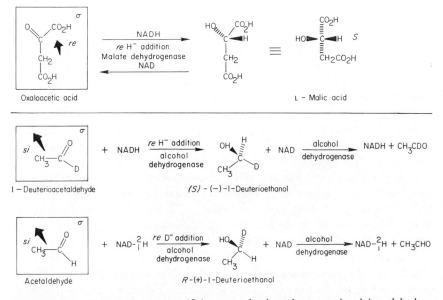

Figure V-32. Biological stereospecificity toward trigonal centers involving dehydrogenases.

addition of hydride to the *si* face of the oxaloacetate ketone group would lead to the production of D-malate [(R)-malate]. (Cf. Figure V-32.)

We have already described the yeast alcohol dehydrogenase reaction as involving a stereospecific transfer of the *pro-R* hydrogen of ethanol to the *pro-R* hydrogen position at C-4' of NADH (p. 77). Biological stereospecificity toward enantiotopic and diastereotopic paired hydrogens was first observed and investigated in this enzymatic reaction. In the early 1950s, soon after Ogston had emphasized how biological differentiation between the paired branches of citrate could occur, it was established that yeast alcohol dehydrogenase differentiated between the paired enantiotopic methylene hydrogen atoms at the C-1 position of ethanol, the paired diastereotopic hydrogens at the C-4' position of the dihydronicotinamide ring of NADH, and the enantiotopic faces of the acetaldehyde carbonyl. It was demonstrated, for example, that two enantiomeric forms of 1-deuterioethanol could be prepared by performing the enzymatic reduction of acetaldehyde in two contrasting ways. Loewus et al. found that the 1-deuterioethanol product formed by the alcohol dehydrogenase-catalyzed reduction of 1-deuterioacetaldehyde by NADH was distinguishable from the 1-deuterioethanol product formed by the enzymatic reduction of unlabeled acetaldehyde by NAD-^{2}H (NADH labeled at the C-4' position of the dihydronicotinamide ring with deuterium).[30]

The two products could be clearly differentiated by running the reverse enzymatic reaction, the oxidation of ethanol by NAD. When the 1-deuterio-ethanol samples were oxidized by NAD in the presence of yeast alcohol dehydrogenase, the 1-deuterioethanol formed enzymatically from 1-deuterio-acetaldehyde yielded nondeuterated NADH and 1-deuterioacetaldehyde, and the other 1-deuterioethanol sample yielded NAD-^{2}H (deuterated NADH) and nondeuterated acetaldehyde. Since both original products were 1-deuterioethanols, they could only differ in stereochemistry at the C-1 position. The two 1-deuterioethanol samples were thus established as enan-tiomers by their contrasting enzymatic oxidation products.

The enantiomeric nature of the 1-deuterioethanols was further sub-stantiated when Loewus et al. chemically converted the stereoisomer origi-nally produced by enzymatic reduction of 1-deuterioacetaldehyde to 1-deuterioethyl tosylate, hydrolyzed the tosylate, and recovered the 1-deuterio-ethanol.[30] This chemical sequence was known to involve an inversion at the C-1 carbon of the ethyl moiety, and therefore the reisolated 1-deuterio-ethanol was the enantiomer of the original starting alcohol. As expected, this reisolated 1-deuterioethanol now reacted like the 1-deuterioethanol produced by NAD-^{2}H reduction of acetaldehyde, and upon enzymatic oxidation it yielded NAD-^{2}H and nondeuterated acetaldehyde.

These examples of biological stereospecificity manifest in the dehydro-genase catalyzed reductions of the trigonal *pro*-chiral carbonyl group of acetaldehyde are illustrated in Figure V-32.

Although the results of Loweus et al. established that the yeast alcohol dehydrogenase reaction displays stereospecificity toward the *pro*-chiral trigonal center of acetaldehyde during reductions and toward the *pro*-chiral methylene group of ethanol during oxidation, these results did not establish the *absolute stereochemical* course of these processes (outlined in Figure V-32). The absolute stereochemical course of this reaction was established by deter-mining the configuration of the 1-deuterioethanol product formed by enzymatic reduction of 1-deuterioacetaldehyde. The 1-deuterioethanol from a large-scale enzymatic reaction was isolated by Levy et al. and found to be levorotatory.[31] Initially it was believed that this levorotatory 1-deuterio-ethanol possessed the R configuration, since stereospecific chemical synthesis of (S)-1-deuterio-1-butanol yielded a dextrorotatory alcohol product.[32] Later results, however, indicated that the deuterium-labeled ethanols and butanols of corresponding absolute configurations rotated plane-polarized light in opposite directions.[33] Finally, a definitive stereospecific synthesis of (R)-1-deuterioethanol was carried out and the R sample found to be dextrorota-tory.[34] The levorotatory 1-deuterioethanol formed by NADH reduction of 1-deuterioacetaldehyde in the presence of yeast alcohol dehydrogenase is therefore (S)-1-deuterioethanol, as shown in Figure V-32. We see that this result requires the dehydrogenase-catalyzed formation of ethanol to involve the stereospecific transfer of hydride from NADH onto the *re* face of the trigonal aldehyde carbon of acetaldehyde. This result also requires that the *pro-R* hydrogen of ethanol be transferred to NAD during the stereospecific dehydrogenase-catalyzed ethanol oxidation. *

It can be readily understood that the production of an alcohol group with the *S* absolute configuration in the malate dehydrogenase reaction is related to the potential formation of diastereomeric complexes between the enantiotopic faces of oxaloacetate and the dissymmetric enzyme. The combination of the enantiomeric (*R*)-malate with the same dissymmetric dehydrogenase catalyst would clearly be diastereomeric to that of the (*S*)-malate combination. Thus the potential *re* and *si* reaction pathways for oxaloacetate reduction will involve diastereomeric interactions and will be of different energies. The citrate synthetase and alcohol dehydrogenase reactions are dependent on the same fundamental principles of differentiation. In the latter examples of biological stereospecificity toward trigonal *pro*-chiral groups, however, the trigonal *pro*-chiral center is converted into a saturated *pro*-chiral center. This *incidental fact* tends to obscure the existence of the close analogies with the malate dehydrogenase reaction. Nevertheless, the two carboxymethyl branches attached to the *pro*-chiral C-3 atom of citrate, and the two methylene hydrogens at C-1 of ethanol, are enantiotopic paired groups. The stereospecific conversion of the attacking reagent (acetyl CoA in the citrate synthetase reaction and hydride in the alcohol dehydrogenase reaction) uniquely into one of the paired groups of the resulting saturated *pro*-chiral center will involve interaction pathways that are diastereomeric with respect to the pathways that would convert the attacking reagents into the enantiotopic paired groups.

LACK OF BIOLOGICAL STEREOSPECIFICITY TOWARDS AN X_{aab} TRIGONAL CENTER: GLYCEROL METABOLISM VIA DIHYDROXYACETONE

This book emphasizes the widespread phenomenon of biological stereospecificity between chemically like, paired groups. Indeed, the lack of such stereospecificity in a reaction may be taken as evidence that the reaction is not enzymatically catalyzed. On the other hand, lack of *overall stereospecificity in a biochemical sequence* may well reflect an intermediate with equivalent, nondifferentiable, paired groups. Once the principles underlying the phenome-

---

* The stereospecificity displayed by the alcohol dehydrogenase reaction with regard to the paired hydrogens at C-4' of the dihydronicotinamide ring of NADH (*pro-R* hydrogen of NADH is transferred) is discussed in detail in Chapter VI (pp. 240–244).

non of biological stereospecificity are clearly understood, an observed *lack of stereospecificity* can be used to deduce useful information about a biochemical sequence.

We have previously described the experiments of Karnovsky et al. with specifically labeled glycerol molecules[6] (pp. 115–116). When incubated with liver slices, the various labeled glycerol molecules were stereospecifically converted into L-α-glycerol phosphate by the glycerol kinase reaction. In contrast, when Rush et al. used the same stereospecifically labeled glycerol molecules (cf. Figure V-4) to study metabolism of glycerol by two *Aerobacter aerogenes* strains, only one *A. aerogenes* strain metabolized the labeled glycerols stereospecifically to produce differentially labeled pyruvate-^{14}C products.[35] When the specifically labeled glycerol substrates were metabolized by *A. aerogenes* strain 1033, pyruvate nearly equally labeled in the 1 and 3 positions was isolated in all cases. Pyruvate-1,3-^{14}C was obtained from *A. aerogenes* (1033) metabolism regardless of whether the glycerol substrate had been synthesized from DL-serine-1-^{14}C (glycerol-1,3-^{14}C), from D-serine-1-^{14}C [(R)-glycerol-1-^{14}C], or from L-serine-1-^{14}C [(S)-glycerol-1-^{14}C] (cf. Figure V-4). As in the case of the citrate synthetases, it is possible that the glycerol kinase enzymes from different sources would have *different stereospecificities*. However, the results just cited indicate a total *lack of metabolic discrimination by the A. aerogenes 1033* strain between the geometrically distinct *pro-R* and *pro-S* branches of the glycerol substrates. This observation is not consistent with the biological metabolism of glycerol via glycerol phosphate. Glycerol, however, may also be converted by glycerol dehydrogenase (Glycerol:NAD oxidioreductase) to dihydroxyacetone as illustrated in Figure V-33. The dihydroxyacetone molecule is an example of an X_{aab} trigonal center which possesses a C_2 axis. This axis permits the rotational superpositioning of the two (a) groups. The two primary alcohol groups of dihydroxyacetone are therefore equivalent paired groups and no biological differentiation between them is possible. Conversion of either the (R) or (S)-glycerol-1-^{14}C substrate to dihydroxyacetone will thus lead to randomization of the original stereospecific label between the terminal carbon atoms. The observed lack of stereospecificity during the conversion of glycerol to pyruvate by *A. aerogenes*

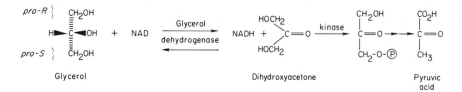

Figure V-33.   Metabolism of glycerol via dihydroxyacetone.

(1033) therefore led to the conclusion that glycerol metabolism proceeds through dihydroxyacetone in this organism.[35]

It is also instructive to note that the C_2 rotational axis of the dihydroxyacetone molecule lies in the plane of the trigonal group. The C_2 rotation therefore interchanges the two faces of the X_{aab} trigonal grouping. Hence differentiable *re* and *si* faces of the dihydroxyacetone carbonyl group do not exist. Addition of hydride ion to either of the equivalent faces of the carbonyl function by the glycerol dehydrogenase enzyme leads to the *same glycerol molecule*. This example illustrates that the possibility of converting a trigonal center to a saturated *pro*-chiral center (C-2 of glycerol) by addition does not, in itself, mean that the unsaturated trigonal center is *pro*-chiral (cf. previous examples of oxaloacetate and acetaldehyde.)

To this point we have emphasized the vital, indeed, determining role that substrate symmetry plays in the phenomenon of biological stereospecificity between chemically like, paired substituents. Purposely, the kind of pictorial representations of three-point enzyme-substrate attachments commonly invoked in biochemical texts to "explain" this phenomenon have been used only to document the historical development of the concept of biological stereospecificity. When utilized, the three-point enzyme-substrate interaction has been described as a convenient representation, which only illustrates how the geometrically nonequivalent groups of a substrate may be differentiated. Before proceeding in Chapter VI with further examples of reactions that display stereospecificity between paired groups, however, it is appropriate to describe an example of a substrate-enzyme interaction that demonstrates how, in reality, biological stereospecificity becomes manifest.

BIOLOGICAL STEREOSPECIFICITIES RESULTING FROM PHYSICALLY REAL SUBSTRATE-ENZYME ATTACHMENTS: SUBSTRATE–α–CHYMOTRYPSIN INTERACTIONS

The most direct way to establish the stereochemistry of a substrate-enzyme interaction would be to obtain and interpret a high-resolution X-ray diffraction pattern from such an interaction. However, in a recent review on X-ray diffraction studies of enzymes, Blow and Steitz note that

The mode of substrate binding has been more difficult to deduce from the three-dimensional structure of the native enzyme than might have been expected. In no case has it been correctly deduced from the *native enzyme structure alone*,* although the prominent clefts in lysozyme and ribonuclease point clearly to the general region in which the substrate must lie.[36]

* My italics.

At the same time, X-ray diffraction data from a substrate-enzyme combination are experimentally difficult to obtain and require careful interpretation before the picture of an active substrate-enzyme interaction can be deduced. Limitations of X-ray diffraction technique* dictate that structure determinations can only be made of states that are stable for the order of one day. Since this is much slower than the catalytic performance of any biologically active enzyme, various artifices must be adopted in order to determine the structures of substrate-enzyme complexes. Therefore, the X-ray diffraction technique cannot directly establish the mode of binding of a natural (reactive) substrate to a native enzyme under optimal catalytic conditions.

Although the technique of X-ray diffraction has already supplied important information regarding a few substrate-enzyme combinations and will undoubtedly make many significant contributions in this area in the future, the technique is not a panacea. Quoting again from the enzyme X-ray crystallographers Blow and Steitz

> The unique features of X-ray diffraction are the detail, precision, and certainty with which structure can be determined in favorable cases. It is, however, no more than a technique, and is limited in its application. It would be absurd to consider diffraction results in isolation from the other techniques of enzymology.[37]

Even in the determination of what is essentially a structural problem, the stereochemistry of a catalytically active substrate-enzyme interaction, therefore, X-ray diffraction data will supplement, not supplant, the data obtained by other techniques.

The study of α-chymotrypsin is a clear example of the fruitful merging of various experimental approaches to the problem of enzymatic activity and stereospecificity. α-Chymotrypsin is a digestive enzyme originating in the pancreas. It is a proteolytic enzyme that catalyzes the hydrolysis of peptide (amide) and of ester bonds. Our knowledge of the stereochemistry of substrate–α-chymotrypsin interactions and of the α-chymotrypsin-catalyzed reaction process is based on an extensive series of kinetic studies,[38–44] special chemical reactions of the catalytic site, [45–49] and, most recently, X-ray diffraction studies.[50,51] X-Ray investigations have also increased our understanding of the process of activation of the zymogen, chymotrypsinogen.[52] This combination of experimental approaches has resulted in a detailed understanding of those aspects of substrate-α-chymotrypsin interactions which are involved in determining the stereospecificity of α-chymotrypsin catalysis. This example, therefore, is ideally suited for our special purpose— demonstrating how biological stereospecificity becomes manifest in an enzymatically catalyzed reaction. The interpretations of substrate-α-chymotrypsin associations proposed by Cohen and his co-workers[53] are particularly

* These limitations are also discussed by Blow and Steitz; cf. Ref. 36, pp. 64–68.

well suited for this special purpose. The following description is based on the analysis developed by Cohen.

In common with other enzymes, α-chymotrypsin displays both reaction specificity and substrate specificity. The particular reaction(s) catalyzed by the enzyme determines its overall reaction specificity, whereas the relative rates of the reactions with various active substrates determine the substrate specificity. In the case of α-chymotrypsin, the observation that the hydrolysis of both amide and ester bonds is catalyzed establishes the reaction specificity. The catalyzed hydrolysis of esters is more rapid, and esters have been utilized as substrates in many experiments designed to explore the precise requirements for the association and for the hydrolysis processes. These experimentally determined requirements have established the substrate specificity of α-chymotrypsin. It is helpful to consider this substrate specificity in terms of two factors, a steric specificity and a relative reaction specificity.*

Steric specificity toward a substrate results when only one enantiomeric form of a dissymmetric substrate (D or L) may bind effectively at the catalytic site. There is lack of steric specificity when either enantiomeric form of the substrate may bind effectively.

The relative reaction specificity of α-chymotrypsin toward a series of substrates is most accurately represented by the values of the specificity constant, defined as the experimentally determined k_{cat}/K_m ratios.[42,54] The k_{cat} value for an active substrate is the unimolecular catalytic rate constant and reflects the relative efficiency of the catalysis processes. (The larger the k_{cat}, the more rapid the catalyzed hydrolysis). The K_m value is the apparent equilibrium constant for dissociation of the enzyme-substrate complex and reflects the effectiveness of the binding of an active substrate to the enzyme (the smaller the K_m, the more effective the enzyme-substrate association). The larger the value of the specificity constant, k_{cat}/K_m, for a given substrate, therefore, the greater the specificity that exists for α-chymotrypsin to catalyze the hydrolysis of the particular structure presented by the substrate.

Studies of the specificity of α-chymotrypsin have shown that the enzyme has a marked preference for substrates with β-aryl (aromatic) residues. It also shows a steric specificity for substrates of the L absolute configuration at the center that is alpha to the acyl function undergoing hydrolysis. The peptide bonds of a protein are therefore preferentially hydrolyzed by α-chymotrypsin at positions where the acyl function is derived from one of the natural aromatic amino acids, L-phenylalanine, L-tyrosine, or L-tryptophan. The methyl N-acetylphenylalaninate (MAPhe) pictured in Figure V-34

---

* The latter specificity factor, the relative reactivity of substrates, should not be confused with the overall reaction specificity of α-chymotrypsin, mentioned previously, which determines what *classes* of compounds may potentially serve as substrates for the enzyme.

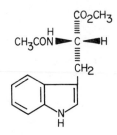

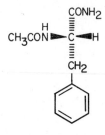

Methyl L–N–Acetyltryptophanate

$(k_{cat}/K_m = 29.2 \times 10^4 m^{-1} sec^{-1})$

L–N–Acetylphenylalaninamide

$(k_{cat}/K_m = 1\ m^{-1}\ sec^{-1})$

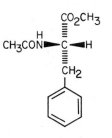

L – MAPhe

Methyl L–N–Acetylphenylalaninate

$(k_{cat}/K_m = 6.2 \times 10^4 m^{-1} sec^{-1})$

Figure V-34.   Effective substrates for α-chymotrypsin-catalyzed hydrolyses.

may be considered to be a "natural" ester substrate for α-chymotrypsin. This ester substrate is hydrolyzed with L steric specificity and with a very high relative reaction specificity $(k_{cat}/K_m = 6.2 \times 10^4\ M^{-1}\ sec^{-1})$.

Table V-2 summarizes the two aspects of substrate specificity for a series of propionate esters in the α-chymotrypsin-catalyzed hydrolysis reaction. Considering the relative reaction specificity constants, it can be seen that the high reactivity of MAPhe is primarily attributable to the β-phenyl substituent. The analogous compound lacking this β-phenyl group, methyl N-acetylalaninate [CH$_3$CH(NHCOCH$_3$)CO$_2$CH$_3$], has a low specificity constant. It is also known that many simple uncharged aromatic compounds bind tightly to α-chymotrypsin and thereby act as inhibitors. We conclude, therefore, that one of the primary binding and orienting sites of α-chymotrypsin is a hydrophobic pocket *ar* which accepts the β-aryl substituent of the active amino acid ester or amide substrate.

Table V-2   Representative Substrate Specificities for α-Chymotrypsin-Catalyzed Hydrolyses[53]

Substrate	Steric specificity	Reaction specificity, k_{cat}/K_m (M^{-1} sec^{-1})
$C_6H_5CH_2CH(NHCOCH_3)CO_2CH_3$ (MAPhe)	L	6.2×10^4
$CH_3CH(NHCOCH_3)CO_2CH_3$ (Methyl N-acetylalaninate)	L	2
$C_6H_5CH_2CH(H)CO_2C_2H_5$ (Ethyl β-phenylpropionate)	—	15
$C_6H_5CH_2CH(OCOCH_3)CO_2C_2H_5$ (Ethyl β-phenyl-α-acetoxypropionate)	L	26
$C_6H_5CH_2CH(CH_3)CO_2C_2H_5$ (Ethyl β-phenyl-α-methylpropionate)	L	3
$C_6H_5CH_2CH(CH_2CO_2C_2H_5)CO_2C_2H_5$ (Ethyl β-phenyl-α-carbethoxymethylene-propionate)	L	8
$C_6H_5CH_2CH(OH)CO_2C_2H_5$ (Ethyl β-phenyl-α-hydroxypropionate)	L and D	1×10^3 11
$C_6H_5CH_2CH(Cl)CO_2C_2H_5$ (Ethyl β-phenyl-α-chloropropionate)	L and D	35 35

Table V-2 indicates that the α-acetylamino group of MAPhe is also required for the high reactivity. Ethyl β-phenylpropionate, which lacks this group, displays a low reaction specificity constant. Furthermore, ethyl β-phenyl-α-acetoxypropionate, in which the NH of the acetylamino group is replaced by an oxygen atom, ethyl β-phenyl-α-methylpropionate, in which the acetylamino group is replaced by methyl, or ethyl β-phenyl-α-carbethoxy-methylenepropionate, in which the acetylamino group is replaced by a carbethoxymethylene group, all display low reactivity in this enzymatic system. We thus conclude that a second binding and orienting site *am* of the α-chymotrypsin provides a hydrogen bonding association with the NH of the α-acylamino group of an active substrate.

Evidence for yet another orienting site of α-chymotrypsin is found in the steric specificities toward active substrates summarized in Table V-2.

α-Chymotrypsin specifically hydrolyzes the L enantiomers of both MAPhe and ethyl β-phenyl-α-methylpropionate. When, however, the α substituent of a β-phenylpropionate substrate is smaller than methyl, as in ethyl β-phenyl-α-hydroxypropionate and in ethyl β-phenyl-α-chloropropionate, the results in Table V-2 indicate that catalyzed hydrolysis of both the D and L substrate enantiomers occurs. Indeed, in the case of the α-chloro substrate, both enantiomers display the *same* low reaction specificity constant. To explain these observations, we propose that a third orienting site of restricted volume h exists which accepts the α-hydrogen atom when the α-acylamino group of an active L substrate binds to site am. In a substrate such as ethyl β-phenyl-α-methylpropionate, the low specificity constant reflects the failure of the α-methyl group to be effectively bonded at the acylamino site am. Nevertheless, L steric specificity is observed because the restricted volume site h, which normally accepts the α-hydrogen of an L enantiomer, cannot accommodate the α-methyl group. Thus effective binding of the D enantiomer, which would require the α-methyl substituent at site h and the α-hydrogen at site am, is prevented. In the case of the substrate structure presented by ethyl β-phenyl-α-chloropropionate, however, the chlorine atom can be accommodated within the volume of the h site. This permits the D and L enantiomers to bind equally effectively to the α-chymotrypsin active site and results in the loss of steric specificity.

Finally, of course, hydrolysis by α-chymotrypsin requires a fourth site n, which is the region containing the catalytically active serine hydroxyl and histidine imidazolyl groups.[55] Effective binding of an active substrate to α-chymotrypsin requires that the acyl group undergoing hydrolysis be situated within this n site.

Based on data such as are represented in Table V-2, Cohen proposes that the active region of α-chymotrypsin may be differentiated into these four defined sites—ar, am, h, and n—with the relative arrangement depicted in Figure V-35A.

The effective association of an active substrate with α-chymotrypsin depicted in Figure V-35B begins with the placement of the β-aryl group in the hydrophobic ar cavity. This primary association leads to an activating conformational change of the enzyme structure. To be effective, the association must also place the acyl group to be hydrolyzed within the catalytic n site. Secondary orienting factors then involve hydrogen bonding of the α-acylamino group at the am site with subsequent placement of the α-hydrogen atom in the restricted volume site h. The hydrogen bonding association of the α-acetylamino and α-hydroxyl groups of L-MAPhe and methyl L-β-phenyl-α-hydroxypropionate at am lead to high and moderate reactivities, respectively. The L-α-acetoxyl, L-α-carbethoxymethyl, L-α-methyl, and L-α-

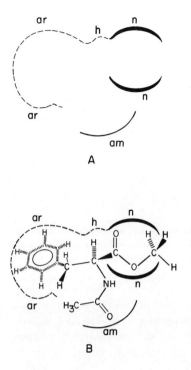

Figure V-35. (A) Active region of α-chymotrypsin according to formulation of S. G. Cohen.[53] (*ar* = hydrophobic pocket binding β-aryl moiety; *am* = binding region of α-acylamino function; *h* = restricted-volume region for α-hydrogen of L substrates; *n* = catalytically active site for acyl group undergoing hydrolysis).
(B) Effective association of methyl L-(+)-N-acetylphenylalaninate with active region of α-chymotrypsin.

chloro groups may also be placed within the *am* site, but the failure of these substituents to hydrogen bond effectively to *am* results in lower reaction specificity constants for such substrates. L Steric specificity is manifest by these substrates when the α-substituting group cannot be accommodated within the small *h* site. Even an α-methyl substituent is too large to be accommodated at the *h* site, and only the L enantiomer of ethyl β-phenyl-α-methylpropionate is hydrolyzed. When the β-phenylpropionate substrates are substituted at the α-position with the smaller chloro or hydroxyl groups, however, these groups may be accommodated at either the *am* or the *h* site. As illustrated in Figure V-36 this permits an effective association of both D and L enantiomers within the active region, and strict steric specificity for L enantiomers is lost.

The relative sizes of the *h* and *am* sites and the hydrogen bonding feature at the *am* site, therefore, largely determine the steric specificity of the reaction. This interpretation is consistent with the observation that the D and L-β-phenyl-α-hydroxylpropionates, with α-hydroxyls that can be sterically accommodated at either *h* or *am*, but can hydrogen bond only at *am*, are both hydrolyzed, but with the L enantiomer showing a higher reactivity. The

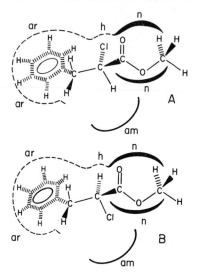

Figure V-36.   Effective association of D (A) and L (B) β-phenyl-α-chloropropionates with the active region of α-chymotrypsin.

substrate derived by replacing the α-hydrogen atom of MAPhe with a methyl group, methyl N-acetyl-α-methylphenylalaninate $[C_6H_5CH_2C(CH_3)-(NHCOCH_3)CO_2CH_3]$, is essentially inactive in the enzymatic hydrolysis reaction, which also emphasizes the role of the restrictive volume h site in determining the substrate specificity.

In addition to the features of substrate specificity just described, α-chymotrypsin displays stereospecificity toward the chemically like, paired substituents at *pro*-chiral centers of appropriate molecules. This aspect of the biological stereospecificity of α-chymotrypsin is also readily explained by the arrangement of binding, orienting, and catalytic sites proposed in Figure V-35A.

α-Chymotrypsin catalyzes the hydrolysis of one of the chemically paired ester functions of the diesters of malonic acid $CH_2(CO_2R)_2$ to yield half-acid esters. In suitably substituted malonates [i.e., $CHR(CO_2R')_2$], the molecules possess a *pro*-chiral center and the catalyzed hydrolysis yields the optically active half-acid esters resulting from a stereoselective reaction at one of the geometrically distinct ester groups.

Diethyl α-acetylaminomalonate $[CH_3CONHCH(CO_2C_2H_5)_2]$ and diethyl α-acetoxymalonate $[CH_3CO_2CH(CO_2C_2H_5)_2]$ are two appropriate substrates which undergo such stereospecific hydrolysis reactions when acted on by α-chymotrypsin. It is proposed that the association of these malonic ester substrates with α-chymotrypsin resembles that of the β-phenylpropionate substrates. The acetylamino or acetoxy groups of these malonate substrates can associate with the *am* site, whereas the α-hydrogen atom is

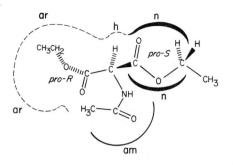

Figure V-37. Binding of diethyl α-acetylaminomalonate to α-chymotrypsin to produce a stereospecific ester hydrolysis.

accommodated at h. As illustrated in Figure V-37, this places one of the ester groups in the catalytic site n and forces the paired ester group to occupy the ar cavity. An anticipated, the lack of an appropriate aryl group on these substrates which could bind effectively at ar results in low specificity constants (k_{cat}/K_m for diethyl α-acetylaminomalonate = 23). That these malonate substrates lack rotational symmetry, however, is of greater stereochemical significance. This dictates, by Hirschmann's superpositioning test, that the two paired ester functions cannot be interchanged if the α-hydrogen and the α-acetylamino groups are to remain associated with their h and am sites, respectively. The enzyme-substrate binding of Figure V-37, therefore, uniquely positions the *pro-S* paired ester function in the hydrolytic site n of α-chymotrypsin. For this reason the enzymatically catalyzed hydrolysis of the *pro*-chiral diethyl malonates is generally stereospecific and only one enantiomeric form of the chiral half-acid ester product is formed.

Diethyl α-hydroxymalonate [$CHOH(CO_2R)_2$] is also a *pro*-chiral molecule. Hydrolysis of this malonate substrate by α-chymotrypsin, however, leads to a racemic half-acid ester product. The rationalization for this contrasting observation appears in Figure V-38. As described previously, an α-hydroxyl group may be accommodated at either the h or the am site. Figure V-38 illustrates these two possible binding modes for the diethyl α-hydroxymalonate substrate. Although this *pro*-chiral molecule lacks rotational symmetry and thus possesses geometrically distinct ester functions, the α-hydroxyl group may be positioned at either h or am, and this allows both the *pro-R* and the *pro-S* ester groups to be placed within the catalytically active site n. *Only one ester function of each molecule is hydrolyzed by α-chymotrypsin and each product is a chiral half-acid ester. However, the enzyme can no longer differentiate between the two geometrically nonequivalent groups.* This leads to a randomly catalyzed hydrolysis of different ester functions on different molecules, and the total half-acid ester product is thus a racemic mixture. Diethyl α-hydroxymalonate, therefore, is an example of a molecule that contains a

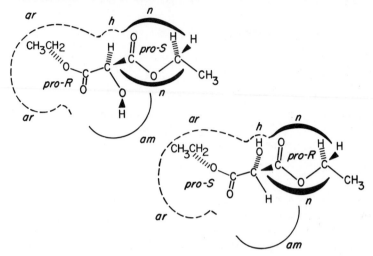

Figure V-38.   Possible effective binding modes of diethyl α-hydroxymalonate.

reactive *pro*-chiral group but which, nevertheless, fails to undergo an enzymatically catalyzed reaction in a stereospecific manner. This *pro*-chiral substrate has been modified to the point that the chemically paired, geometrically nonequivalent groups cannot be differentiated by the restricting and orienting sites of the α-chymotrypsin enzyme.

A third class of ester molecules may be used to demonstrate some additional stereochemical features of α-chymotrypsin–substrate associations. In general those active substrates which are hydrolyzed by α-chymotrypsin without total steric specificity for the L configuration (either no stereospecificity or preference for the D configuration) have also been found to display low reaction specificity constants ranging from 0.1 to 65 M^{-1} sec^{-1}. A type of ester substrate was discovered by Niemann and his co-workers, however, which displayed D steric specificity coupled with a high reaction specificity ($k_{cat}/K_m = 4.3 \times 10^4$ M^{-1} sec^{-1}) comparable to that of the "natural" ester substrate, L-MAPhe.[56] These substrates are cyclized structures and may be represented by the 3-methoxycarbonyl-3,4-dihydroisocarbostyril (MCIC) molecule, illustrated in its D and L enantiomeric forms in Figure V-39.

Initially, we may consider the MCIC structure to be merely an analog of MAPhe in which the α-acylamino group has been fused to the β-aryl ring. A reasonable prediction, therefore, would be that the aromatic ring of MCIC would bind to *ar*, the lactam function (cyclized amide group) to *am*, with the methoxycarbonyl group (ester function) placed at the *n* site. However,

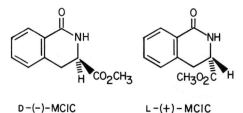

D–(–)–MCIC   L–(+)–MCIC

Figure V-39.   D and L Enantiomers of MCIC.

such a straightforward interpretation, based on analogy with the formulated view of α-chymotrypsin-MAPhe association, is not consistent with the D steric substrate specificity displayed by the cyclized MCIC molecule. On closer examination, it becomes apparent that the fused lactam ring of L-MCIC produces an overall molecular geometry which prevents binding in a manner directly analogous to that proposed for L-MAPhe (see Figure V-40). As shown, when the ester function is placed at the catalytic *n* site and the L-α-lactam group is situated within the *am* site, the aromatic ring cannot be positioned in the *ar* cavity.

   In contrast, the high reaction specificity of the cyclized D-MCIC enantiomer would seem to reflect the fact that this enantiomer *can* be positioned with the aromatic ring at the primary *ar* binding site and the ester function simultaneously positioned at the catalytic site *n*. This interpretation is made more plausible when it is realized that the lactam ring produces a molecular geometry such that the fused aromatic ring *and* the lactam group may both be accommodated within the combined *ar–h* region. This proposed effective binding mode for D-MCIC with the active region of α-chymotrypsin is depicted in Figure V-41. *Such an association is not possible in the case of a substrate possessing a noncyclized D-acylamino group.*

   Figure V-41 illustrates clearly how the high reaction specificity constant

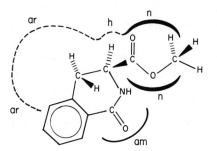

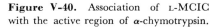

Figure V-40. Association of L-MCIC with the active region of α-chymotrypsin.

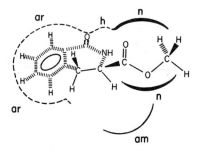

Figure V-41. Effective association of D-MCIC with the active region of α-chymotrypsin.

for α-chymotrypsin-catalyzed hydrolysis of D-MCIC can be determined by binding of the aromatic ring at the *ar* site rather than by the binding of the lactam function at the *am* site (cf. Figure V-40). Indeed, in the proposed association of Figure V-41, no specific use is made of the lactam NH in either orienting or binding the molecule. The validity of the interpretation of D-MCIC reactivity schematically depicted in Figure V-41 is therefore established by the rapid enzymatic hydrolysis ($k_{cat}/K_m = 6.0 \times 10^4 \ M^{-1} \sec^{-1}$) of the D enantiomer of the cyclized substrate in which the NH of D-MCIC has been replaced by an oxygen atom, D-3-methoxycarbonyl-3,4- dihydroisocoumarin.[57] This substrate analog of D-MCIC does not possess an amide (or lactam) function, nor is the D enantiomer of its close, noncyclized counterpart, ethyl β-phenyl-α-acetoxypropionate, hydrolyzed by α-chymotrypsin (Table V-2).

The substituted dihydroisocarbostyril and dihydroisocoumarin molecules thus represent substrate analogs in which the molecular geometry has been modified to the point that the "normal" complementary binding of enzyme and substrate becomes impossible (cf. Figure V-40). At the same time, the resulting change in molecular geometry makes possible a modified but effective mode of association with the primary binding and catalytic sites of the enzyme (Figure V-41). This modified form of enzyme-substrate association is "unnatural" only because it is not available to the noncyclized substrates. The result is a class of highly reactive substrates in which the "natural" steric specificity of α-chymotrypsin for the L configuration has been reversed.

The results of X-ray diffraction studies on the covalently linked acyl-enzyme complex, indoleacryloyl–α-chymotrypsin,[51] and on the reversibly formed N-formyl-L-tryptophan–α-chymotrypsin complex[50] have established a three-dimensional picture of the binding and catalytic sites of α-chymotrypsin which closely resembles the model previously deduced by Cohen. Figure V-42 presents the active site region of α-chymotrypsin, revealed by

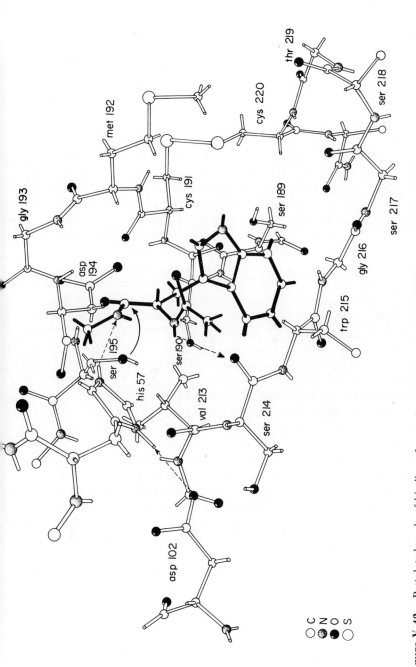

Figure V-42. Postulated mode of binding of an active amide substrate within the active site of α-chymotrypsin. Also indicated is the protonation of the NH of the bond to be cleaved by His-57, the nucleophilic attack by the Ser-195 oxygen on the acyl carbon of the substrate, and the hydrogen bonding of the NH of the L-α-acylamino group of the substrate by the Ser-214 carbonyl oxygen. (Figure based on the results of Steitz et al.[50] and Henderson.[51])

these X-ray diffraction studies. The postulated mode of binding of an active amide substrate within this region is indicated.

The aromatic side chain of the acyl residue (in this case tryptophanyl) positioned for hydrolysis resides in a pocket in the α-chymotrypsin structure near Ser-195. The most prominent interactions are with peptide bonds whose planes are approximately parallel to the plane of the aromatic ring and 3.5 to 4.0 Å away from it. Peptide bonds 190–191 and 191–192 interact above the ring and peptide bonds 215–216 interact below. The side chains of such residues as Ser-190, Val-213, Try-215, and Gly-216 produce additional van der Waals contacts with atoms of the aromatic ring. These contacts are generally "edge on" to the plane of the ring. These interactions correspond to the bonding within the hydrophobic pocket *ar* of Cohen's model. The X-ray diffraction studies also indicated that the carbonyl oxygen of Ser-214 is properly oriented to hydrogen bond with the NH of the L-α-acylamino group of the amide substrate. This interaction corresponds to that postulated at *am* in Cohen's model. Figure V-42 illustrates how these two binding interactions position the acyl group of the substrate within the catalytic region formed by residues Asp-102, His-57, and Ser-195 of α-chymotrypsin. The *n* region of Cohen's model then corresponds to the arrangement of the vital catalytic groups Asp-102, His-57, and Ser-195 established by the X-ray investigations. The restricted volume region *h* of Cohen's model must correspond to the area of the active site region of α-chymotrypsin which contains the Cys-191 and Met-192 residues. Finally, comparison of Figures V-35 and V-42 indicates that the three-dimensional active site region of α-chymotrypsin as revealed by the X-ray diffraction studies generally corresponds to the proposed three-dimensional arrangement of sites *ar*, *am*, *h*, and *n* of Cohen's model. The X-ray results thus substantiate the analysis of the stereochemistry of substrate–α-chymotrypsin associations based on Cohen's model.

By emphasizing certain aspects of the stereochemistry of α-chymotrypsin–substrate associations, we have attempted to document how the particular topics of this book are involved in an actual enzymatic process. We have seen, in three classes of esters, how an analysis of the relative substrate specificities and reactivities provided an indirect method of mapping the active region of α-chymotrypsin and how recent X-ray investigations defined this active region more precisely. More germane to the subject of this book, we have also seen how the active region of α-chymotrypsin is comprised of oriented sites forming a dissymmetric environment that is complementary to the chemical groups and molecular geometries of the reactive enzymatic substrates. In effect, our analysis of the nature of substrate–α-chymotrypsin associations demonstrates how Ogston's "three-point attachment" concept relates to an actual biological situation.

It should be clear from Figures V-35 to V-41 how the same dissymmetric enzymatic environment leads to diastereomeric interactions with both chiral and *pro*-chiral substrates. It is thus demonstrated that biological stereospecificity toward separate dissymmetric molecules and toward the chemically like, nonequivalent paired groups of a single substrate molecule are dependent on identical active site features of α-chymotrypsin. As emphasized in the earlier chapters, the essential difference between these two aspects of biological stereospecificity lies in the different molecular symmetries of the substrates.

Finally, the behavior of certain of the active substrates discussed illustrates that biological stereospecificity is not an absolute and inviolate feature of enzymatic reactions. Biological stereospecificity involves interactions between the geometry of the substrate and that of the total active site region of the enzyme. It is possible to alter the structure of a substrate molecule to permit interactions with the enzyme surface that are catalytically active but lack some of the normal binding and orienting interactions. In such molecular structures the normally restrictive orienting ability of the active region of a particular enzyme may be exceeded, and biological specificity between chemically like, paired groups or between enantiomeric molecules may be lost, decreased, or reversed.

LACK OF THE EXPECTED BIOLOGICAL STEREOSPECIFICITY IN A CHLOROPEROXIDASE REACTION

The recent finding by Kollonitsch et al. that the chloroperoxidase reaction with propenylphosphonic acid substrates does *not* manifest the expected biological stereospecificity[58] emphasizes the latter point. In the presence of hydrogen peroxide and chloride ion, chloroperoxidase preparations catalyze the formation of chlorohydroxyphosphonates from propenylphosphonates, as in Figure V-43. As indicated in the figure, however, the product of the chloroperoxidase-catalyzed reaction from the *trans*-propenylphosphonic acid was a racemic mixture of the *erythro*-1-chloro-2-hydroxypropylphosphonic acids, whereas the product from the *cis*-propenylphosphonic acid was a racemic mixture of the *threo*-1-chloro-2-hydroxypropylphosphonic acids. Thus this enzymatic reaction fails to show the expected biological stereospecificity, and the enzymatic product in each case is a racemic mixture of 1-chloro-2-hydroxypropylphosphonates. Again, although extreme biological stereospecificity is generally observed in enzymatic reactions, this phenomenon is not an essential feature of biologically catalyzed processes.

This unusual example of the enzymatic synthesis of a racemic mixture of chiral products also illustrates in a concrete manner some differences be-

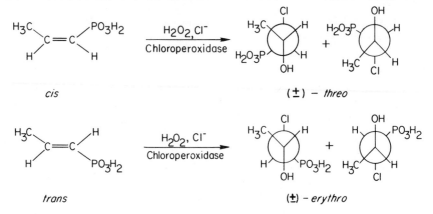

Figure V-43.   Chloroperoxidase reaction with *cis* and *trans*-propenylphosphonic acids.[58]

tween manifestations of *chemical stereospecificity* and *biological stereospecificity* (cf. footnote, p. 5). From the Newman projections of the *erythro* and *threo*-1-chloro-2-hydroxypropylphosphonic acid products in Figure V-43, it is apparent that the formation of these diastereomeric acids from the appropriate propenylphosphonic acids involves a stereospecific *anti* addition of chloride and hydroxide to the double bond of the substrate molecules. In this way, the *cis*-phosphonic acid specifically yields the *threo* stereoisomers when acted upon by chloroperoxidase, and the *trans*-phosphonic acid specifically yields the *erythro* stereoisomers. This behavior is characteristic of *chemical stereospecific* processes.

Normally, we would anticipate that an enzymatic reaction would also display a marked *stereoselectivity* and preferentially catalyze the conversion of one of the two isomeric unsaturated substrates to product. In addition, Figure V-44 shows that these isomeric substrates possess trigonal *pro*-chiral groupings. For example, the *si,si* face of the double bond in *cis*-propenylphosphonic acid is enantiotopically related to the *re,re* face. Therefore, we would expect the chloroperoxidase enzyme to differentiate between these paired faces of the unsaturated *pro*-chiral grouping, and add chloride to either the *si,si* face or the *re,re* face of the *cis*-propenylphosphonic acid. In this manner, normal biological stereospecificity would have led to the production of a single optically active *threo* isomer if the *cis* acid were the reactive substrate, *or* a single optically active *erythro* isomer if the *trans* acid were stereoselectively converted.

The chloroperoxidase-catalyzed reaction described above, therefore, manifests *only* chemical stereospecificity. The process does not evidence normal biological stereospecificity because both isomeric substrates are con-

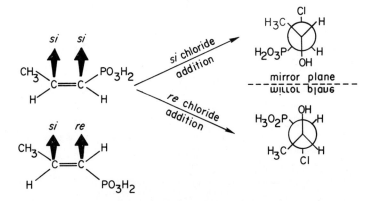

Figure V-44. Potential biologically stereospecific chloroperoxidase reactions.

verted *and* because the enzyme does not additionally differentiate between the paired enantiotopic groups (faces) at the reacting unsaturated *pro-chiral* centers.

REFERENCES

1. A. G. Ogston, *Nature*, **162**, 963 (1948).
2. P. E. Wilcox, *Nature*, **164**, 757 (1949).
3. K. R. Hanson, *J. Amer. Chem. Soc.*, **88**, 2731 (1966).
4. H. Hirschmann, *J. Biol. Chem.*, **235**, 2762 (1960); H. Hirschmann, Chapter VII, in *Comprehensive Biochemistry*, Vol. 12, M. Florkin and E. H. Stotz, Eds., Elsevier, 1964.
5. P. Schwartz and H. E. Carter, *Proc. Natl. Acad. Sci. (U.S.)*, **40**, 499 (1954).
6. M. L. Karnovsky, G. Hauser, and D. Elwyn, *J. Biol. Chem.*, **226**, 881 (1957).
7. S. Aronoff, *Techniques of Radiobiochemistry*, pp. 12–14, Iowa State College Press, 1956.
8. M. H. O'Leary, *J. Amer. Chem. Soc.*, **91**, 6886 (1969).
9. E. A. Evans, Jr., and L. Slotin, *J. Biol. Chem.*, **141**, 439 (1941).
10. H. G. Wood, C. H. Werkman, A. Hemingway, and A. O. Nier, *J. Biol. Chem.*, **139**, 483 (1941).
11. S. Englard, *J. Biol. Chem.*, **235**, 1510 (1960).
12. O. Gawron and T. P. Fondy, *J. Amer. Chem. Soc.*, **81**, 6333 (1959).
13. F. A. L. Anet, *J. Amer. Chem. Soc.*, **82**, 994 (1960).
14. T. C. Farrar, H. S. Gutowsky, R. A. Alberty, and W. G. Miller, *J. Amer. Chem. Soc.*, **79**, 3978 (1957).
15. K. R. Hanson and I. A. Rose, *Proc. Natl. Acad. Sci. (U.S.)*, **50**, 981 (1963).
16. T. Kaneko and H. Katsura, *Chem. Ind. (London)*, 1188 (1960).

17. A. L. Patterson, C. K. Johnson, D. van der Helm, and J. A. Minkin, *J. Amer. Chem. Soc.*, **84**, 309 (1962).

18. D. W. Fanshier, L. K. Gottwald, and E. Kun, *J. Biol. Chem.*, **237**, 3588 (1962); D. W. Fanshier, L. K. Gottwald, and E. Kun, *J. Biol. Chem.*, **239**, 425 (1964); R. J. Dummel and E. Kun, *J. Biol. Chem.*, **244**, 2966 (1969).

19. S. Englard and S. P. Colowick, *J. Biol. Chem.*, **221**, 1019 (1956).

20. T. T. Tchen and H. van Milligan, *J. Amer. Chem. Soc.*, **82**, 4115 (1960).

21. E. J. Wawszkiewicz and H. A. Barker, *J. Biol. Chem.*, **243**, 1948 (1968).

22. R. D. Batt, F. Dickens, and D. H. Williamson, *Biochem. J.*, **77**, 272 (1960).

23. I. A. Rose, E. L. O'Connell, P. Noce, M. F. Utter, H. G. Wood, J. M. Willard, T. G. Cooper, and M. Benziman, *J. Biol. Chem.*, **244**, 6130 (1969).

24. J. P. Glusker, *J. Mol. Biol.*, **38**, 149 (1968).

25. G. Gottschalk and H. A. Barker, *Biochemistry*, **5**, 1125 (1966).

26. G. Gottschalk and H. A. Barker, *Biochemistry*, **6**, 1027 (1967).

27. G. Gottschalk, *Europ. J. Biochem.*, **5**, 346 (1968).

28. J. R. Stern and R. W. O'Brien, *Biochim. Biophys. Acta*, **185**, 239 (1969); R. W. O'Brien and J. R. Stern, *Biochem. Biophys. Res. Commun.*, **34**, 271 (1969).

29. S. Dittbrenner, A. A. Chowdhury, G. Gottschalk, *Biochem. Biophysics Res. Commun.*, **36**, 802 (1969).

30. F. A. Loewus, F. H. Westheimer, and B. Vennesland, *J. Amer. Chem. Soc.*, **75**, 5018 (1953).

31. H. R. Levy, F. A. Loewus, and B. Vennesland, *J. Amer. Chem. Soc.*, **79**, 2949 (1957).

32. A. Streitwieser, Jr., *J. Amer. Chem. Soc.*, **75**, 5014 (1953); *ibid.*, **77**, 1117 (1955).

33. A. Streitwieser, Jr., J. R. Wolfe, Jr., and W. D. Schaeffer, *Tetrahedron*, **6**, p. 341 (1959).

34. R. U. Lemieux and J. Howard, *Can. J. Chem.*, **41**, 308 (1963).

35. D. Rush, D. Karibian, M. L. Karnovsky, and B. Magasanik, *J. Biol. Chem.*, **226**, 891 (1957).

36. D. M. Blow and T. A. Steitz, *Annu. Rev. Biochem.*, **39**, p. 74 (1970).

37. *Loc. cit.*, p. 64.

38. B. S. Hartley and B. A. Kilby, *Biochem. J.*, **56**, 288 (1954).

39. H. Gutfreund and J. M. Sturtevant, *Biochem. J.*, **63**, 656 (1956).

40. L. Faller and J. M. Sturtevant, *J. Biol. Chem.*, **241**, 4825 (1966).

41. C. Niemann, *Science*, **143**, 1287 (1964).

42. M. L. Bender and F. J. Kézdy, *Annu. Rev. Biochem.*, **34**, 49 (1965).

43. A. Himoe and G. P. Hess, *Biochem. Biophys. Res. Commun.*, **27**, 494 (1967).

44. A. Himoe, K. G. Brandt, R. J. DeSa, and G. P. Hess, *J. Biol. Chem.*, **244**, 3483 (1969).

45. A. K. Balls and E. F. Jansen, *Advan. Enzymol.*, **13**, 321 (1952).

46. J. A. Cohen, R. A. Oosterbaan, H. S. Jansz, and F. Berends, *J. Cell. Compar. Physiol.*, **54**, Suppl. 1, 231 (1959).

47. L. Weil, S. James, and A. R. Buchert, *Arch. Biochem. Biophys.*, **46**, 266 (1953).

48. D. E. Koshland, Jr., "The Active Center in Enzyme Action," Brookhaven National Laboratory Publ. 7557, 1963.

49. G. Schoellmann and E. Shaw, *Biochemistry*, **2**, 252 (1963).

50. T. A. Steitz, R. Henderson, and D. M. Blow, *J. Mol. Biol.*, **46**, 337 (1969).

51. R. Henderson, *J. Mol. Biol.*, **54**, 341 (1970).

52. B. W. Matthews, P. B. Sigler, R. Henderson, and D. M. Blow, *Nature*, **214**, 652 (1967); P. B. Sigler, D. M. Blow, B. W. Matthews, and R. Henderson, *J. Mol. Biol.*, **35**, 143 (1968).

53. S. G. Cohen, A. Milovanović, R. M. Schultz, and S. Y. Weinstein, *J. Biol. Chem.*, **244**,, 2664 (1969); S. G. Cohen, *Trans. N.Y. Acad. Sci.*, **31**, 705 (1969).

54. F. E. Brot and M. L. Bender, *J. Amer. Chem. Soc.*, **91**, 7187 (1969).

55. W. P. Jencks, *Catalysis in Chemistry and Enzymology*, p. 218, McGraw-Hill, 1969.

56. G. E. Hein, R. B. McGriff, and C. Niemann, *J. Amer. Chem. Soc.*, **82**, 1830 (1960); G. E. Hein and C. Niemann, *ibid.*, **84**, 4487 (1962).

57. S. G. Cohen and R. M. Schultz, *J. Biol. Chem.*, **243**, 2607 (1968).

58. J. Kollonitsch, S. Marburg, and L. M. Perkins, *J. Amer. Chem. Soc.*, **92**, 4489 (1970).

Chapter VI

ADDITIONAL EXAMPLES OF BIOLOGICAL STEREOSPECIFICITY BETWEEN CHEMICALLY LIKE, PAIRED GROUPS

Biological stereospecificity between chemically like, paired groups is not an esoteric aspect of biochemistry. The discussions in the preceding chapters have illustrated how such selectivity is inherent to the substrate-enzyme interactions that characterize biological processes. This chapter presents selected additional examples of biochemical reactions and processes that manifest biological stereospecificity between chemically like groups. Some of these examples of biological stereospecificity were instrumental in developing an understanding of the total phenomenon. No student should claim a knowledge of biological stereospecificity without some familiarity with these "classical" examples. For this reason, Chapter VI includes analyses of the biological stereospecificities evident in TCA cycle metabolism, in the biosynthetic path to squalene, in the ketol-isomerase reactions, and in the reactions of NADH and NADPH, even though summaries of these fundamental examples of biological stereospecificity may be found in other reviews of the subject. Some of the additional examples, however, although taken from the older scientific literature, have generally been overlooked in other descriptions of biological stereospecificity. The analyses of the biological stereospecificities manifest in the metabolism of the diaminopimelic acids and in the oxidative metabolism of certain inositols fall into this category. These less well-known examples illustrate interesting aspects of biological stereospecificity which were not emphasized in the preceding chapters. Certain of the additional examples presented in Chapter VI show how investigations of biological stereospecificity can yield useful information concerning the mechanisms of enzymatic catalysis. The analyses of δ-aminolevulinic acid synthetase and of the reactions involving the cobalamin (B_{12}) coenzymes, and portions of the analyses of the ketol-isomerase and the aldolase reactions are included for this purpose. Finally, certain of the additional examples of biological stereospecificity between chemically like groups described here are derived from recent experimental investigations which have extended our knowledge of biological stereospecificity in new directions. The stereospecific reactions of chiral acetyl CoA and of chiral pyruvate substrates are such examples.

BIOLOGICAL STEREOSPECIFICITY WITHIN TCA CYCLE METABOLISM: A SUMMARY

The biological stereospecificities manifest in the operations of the TCA cycle are summarized in Figure VI-1.

The biological stereospecificities summarized in Figure VI-1 may be analyzed in terms of the contrasting symmetry properties of three key TCA intermediates—fumarate, citrate, and succinate. Such an analysis affords an excellent opportunity to consolidate several of the important principles developed in earlier chapters. For example, in Chapter V we saw that all the paired chemical groupings of fumarate are superimposable by rotational symmetry operations. As indicated in Figure VI-1, therefore, the paired atoms of fumarate are equivalent and nondifferentiable. Nevertheless, the *pro*-chiral double bond of fumarate possesses geometrically nonequivalent *re,re* and *si,si* faces. As illustrated in Figure VI-1, fumarase catalyzes the addition of hydrogen to the *re,re* face and hydroxyl to the *si,si* face of fumarate to stereospecifically form L-malate. Metabolism through the fumarase and malate dehydrogenase reactions will thus produce an oxaloacetate molecule in which the H_S methylene hydrogen corresponds to one of the nondifferentiable hydrogens of fumarate, whereas the H_R methylene hydrogen represents a proton from the solvent, added in the fumarase reaction.

Evidence for the stereospecific *si* pathway of the normal (*S*)-citrate synthetase was also described in Chapter V (pp. 151–154). At that time, the experiments of Hanson and Rose[1] were used to emphasize the potential *pro*-chiral nature of unsaturated trigonal centers. Figure VI-1 shows that (*S*)-citrate synthetase produces a citrate molecule in which the *pro-S* carboxymethyl branch is derived from the acetyl CoA moiety, whereas the *pro-R* branch, derived from oxaloacetate, bears an H_S methylene hydrogen derived from fumarate and an H_R methylene hydrogen added in the fumarase reaction.

It was emphasized earlier (p. 133) that citric acid lacks rotational symmetry and that all the paired functions must therefore be stereochemically distinct and biologically differentiable. The experiments of England[2] (outlined previously in Figure V-6) established that the hydrogen atom

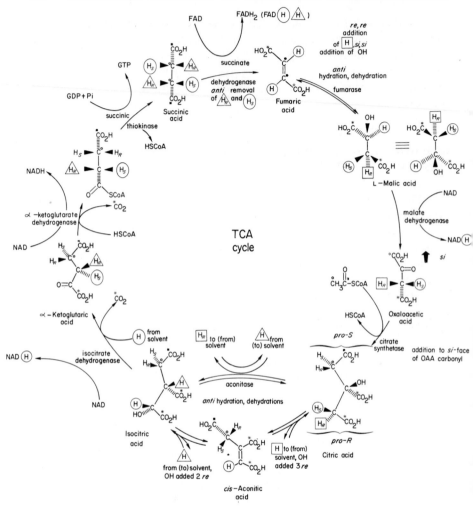

Figure VI-1. Schematic summary of the biological stereospecificities of the TCA cycle.

added in the fumarase reaction is stereospecifically lost in the conversion to isocitrate. As illustrated in Figure VI-1, this observation requires that the H_R hydrogen on the *pro-R* carboxymethyl branch of citrate be removed in the aconitase-catalyzed conversion. The H_S hydrogen present on this *pro-R* branch of citrate is also lost, but this occurs in the oxidative decarboxylation catalyzed by the next enzyme of the cycle. As illustrated, the latter hydrogen atom was originally one of the nondifferentiable hydrogens of fumarate and is removed as NADH.

Although both the methylene hydrogens originally present in oxalo-acetate are stereospecifically removed in the foregoing reactions, the *anti* H_2O addition-elimination reaction catalyzed by aconitase (cf. Figure V-30) and the oxidative decarboxylation reaction catalyzed by isocitric acid dehydro-genase result in the incorporation of two protons from the aqueous media. These two new methylene hydrogens are found on the β-carbon atom of α-ketoglutarate. As indicated in Figure VI-1, the *pro-R* hydrogen at this position represents the proton added in the aconitase reaction; the *pro-S* hydrogen is derived from the solvent in the isocitric dehydrogenase reaction. * The α-ketoglutarate intermediate lacks rotational symmetry and clearly each carbon atom is unique. The stereospecificities of the citrate synthetase reaction and of the aconitase reaction dictate that the carboxyl function adjacent to the keto group of α-ketoglutarate be derived from oxaloacetate, while the γ-methylene group and attached carboxyl function are derived from acetyl CoA. This is, of course, the originally observed Ogston effect in TCA cycle metabolism (cf. pp. 8–14). The next reaction of the cycle, α-ketoglutarate dehydrogenase, thus specifically releases as carbon dioxide the carbon atom originally present as the C-4 carboxyl group of oxaloacetate. The carbon and hydrogen atoms of the resulting succinyl CoA retain their separate chemical and stereochemical identities. Upon conversion to succinate in the succinic thiokinase reaction, however, the situation changes markedly.

As was demonstrated in Figure V-10, Hirschmann's superpositioning test establishes that the chemically paired *carbon* atoms of succinate are superimposable, carboxyl on carboxyl and methylene carbon on methylene carbon. These paired groups are thus equivalent and nondifferentiable. Figure VI-1 reveals that carbon atoms introduced from acetate and from oxaloacetate become randomized at succinate in the TCA cycle. At the same point in the cycle, the methylene hydrogens of succinate, which were origi-nally derived from the methyl hydrogens of acetate, now become super-imposable upon the methylene hydrogens of succinate derived from the solvent in the aconitase and isocitric acid dehydrogenase reactions. Never-theless, as Figure V-10 also demonstrated, the four methylene hydrogens of succinate are not totally equivalent but exist as two pair of enantiotopic atoms. As indicated in Figure VI-1, the *pro-R* hydrogen pair may be repre-sented by the proton added in the aconitase reaction, whereas the *pro-S* pair may be represented by the proton added in the isocitrate dehydrogenase step. In Chapter V (pp. 131–133) we described the elegant experiment of Tchen and van Milligan.[4] This experiment demonstrated that the succinic

---

* It has been established that, as indicated in Figure VI-1, the isocitrate dehydrogenase reaction proceeds with retention of configuration.[3]

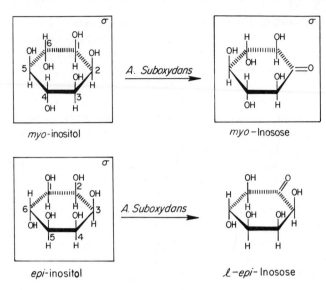

Figure VI-2.   Metabolic transformations of *myo* and *epi*-inositols by *A. suboxydans*.

dehydrogenase reaction removes H_R and H_S hydrogen atoms from the adjacent methylene positions of succinate. Figure VI-1 shows that after the succinic dehydrogenase reaction, the paired chemical groups (including now the hydrogen atoms) have again become equivalent in the fumarate intermediate.

BIOLOGICAL STEREOSPECIFICITIES IN INOSITOL METABOLISM

The known metabolism of some inositol stereoisomers by *Acetobacter suboxydans* affords an excellent opportunity to review and consolidate additional principles of biological stereospecificity developed in earlier chapters. The metabolism of inositols may also be used to illustrate the validity of analyses based on the artificially simplified, planar, cyclohexane representations described in Chapter II (pp. 37–38).

Cultures of *A. suboxydans* have been found to catalyze the oxidation of two of the optically inactive inositol stereoisomers, *myo*-inositol and *epi*-inositol, to monoketones. The structures of the two monoketone products have been established,[5–7] and the processes may be represented by the simplified planar cyclohexane projections of Figure VI-2. Both the inositol isomers in Figure VI-2 are nondissymmetric substrates. They possess reflective symmetry and are optically inactive. In the *myo*-inositol substrate, a mirror plane passes through C-2 and C-5 of the cyclohexane ring. This mirror plane relates the alcohol functions at C-1 and C-3 and at C-6 and C-4 as paired groups. Since the representation in Figure VI-2 indicates that *myo*-inositol lacks all elements of rotational symmetry, it follows that these paired groupings must be enantiotopic paired groups. We see that the oxidative metabolism of *myo*-inositol by *A. suboxydans* involves oxidation of the C-2 hydroxyl to yield the optically inactive *myo*-inosose. This optically inactive monoketone product, however, does not result from a lack of biological stereospecificity in *myo*-inositol metabolism. Rather, the *A. suboxydans* specifically oxidizes the *chemically unique* C-2 hydroxyl group. (Figure VI-2 emphasizes that this is the only hydroxyl of *myo*-inositol that is *cis* to *both* adjacent hydroxyl groups.) Since the chemically unique C-2 hydroxyl oxidized in this biological process lies on the mirror plane of the original *myo*-inositol substrate, the product of this biological *chemically specific* oxidation must be an optically inactive *meso* ketone.

The molecular mirror plane of the *epi*-inositol passes through C-3 and C-6 and relates the alcohol functions at C-1 and C-5 and at C-2 and C-4 as paired groups. Again, owing to the lack of rotational symmetry in the *epi*-inositol stereoisomer, all these paired groupings are enantiotopic. Despite this stereochemical analogy with *myo*-inositol, however, the oxidative metabolism of *epi*-inositol in *A. suboxydans* cultures leads to the production of an optically active monoketone, levorotatory *epi*-inosose. It is apparent from Figure VI-2 that this (*l*)-*epi*-inosose results from the oxidation of the C-2 hydroxyl group of *epi*-inositol. In contrast to the C-2 hydroxyl of *myo*-inositol, the C-2 hydroxyl of *epi*-inositol is enantiotopic to the paired C-4 hydroxyl. The production of the optically active (*l*)-*epi*-inosose, therefore, is the result of a biological stereospecific oxidation which distinguishes between the paired enantiotopic C-2 and C-4 hydroxyl groups of *epi*-inositol on the basis of absolute geometrical nonequivalence.

The oxidative metabolism of the optically active *d-chiro*-inositol and *l-chiro*-inositol substrates in *A. suboxydans* cultures provides yet a third contrasting example. It has been determined that these two enantiomeric substrates are converted into enantiomeric diketone products as indicated in Figure VI-3.[7] This, to be sure, is unusual. Only one of the enantiomeric *chiro*-inositols should have been metabolized according to our expectations; or at least these two stereoisomers should have been metabolized differently (cf. the foregoing examples of *myo* and *epi*-inositol stereoisomers). Quite apart from this unusual feature, however, the metabolism of *d* and *l-chiro*-inositols by the *A. suboxydans* cultures provides an instructive contrast to the metabolism of *myo* and *epi*-inositols. Examination of the simplified planar representations of Figure VI-3 makes it apparent that the *d* and *l-chiro*-inositol stereoisomers, although lacking reflective symmetry (and thus being

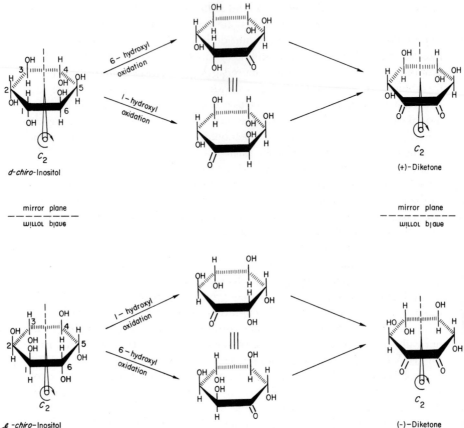

Figure VI-3. Metabolic transformations of *d-chiro* and *l-chiro*-inositols by *A. suboxydans*.

dissymmetric), nevertheless possess rotational symmetry. As illustrated, C_2 rotational axes in the planes of the representations bisect the C-3–C-4 and C-1–C-6 carbon–carbon bonds of the *chiro*-inositols. The C_2 symmetry operations result in the rotational superpositioning of the C-1 and C-6, C-2 and C-5, and C-3 and C-4 hydroxyl groups of these inositols. These paired chemical groups within each molecule are therefore equivalent and are not capable of being differentiated. As Figure VI-3 reveals, the optically active diketones isolated after *chiro*-inositol metabolism results from oxidation of the C-1 and C-6 hydroxyl functions. These metabolic oxidations therefore are seen to involve the oxidation of two equivalent paired functional groups of each substrate molecule. The geometrical equivalence of the paired groups oxidized has been emphasized in Figure VI-3 by illustrating the monoketone structures probably involved as intermediates in the total process. Again, the planar representations help make it apparent that the two monoketone intermediates potentially formed in each separate sequence are in fact equivalent. In the oxidative metabolism, it clearly cannot matter whether the C-1 hydroxyl is oxidized before the equivalent C-6 hydroxyl or vice versa.

These interesting instances of inositol metabolism may be summarized as follows: The metabolism of the nondissymmetric *myo*-inositol by *A. suboxydans* involves the oxidation of a chemically unique hydroxyl function of the substrate lying on the molecular mirror plane and thus yields an optically inactive *meso* product. The metabolism of the nondissymmetric *epi*-inositol by this organism involves the biologically stereospecific oxidation of one enantiotopic paired hydroxyl function of the substrate and thus yields an optically active product. In contrast to the enantiotopic relationships existing between the hydroxyls in the *myo* and *epi*-inositol molecules, the rotational symmetry of the *d* and *l-chiro*-inositol structures creates sets of equivalent paired hydroxyl groups within each of these enantiomeric molecules. The metabolism of the *d* and *l-chiro*-inositol substrates by *A. suboxydans* then involves the oxidation of a pair of equivalent hydroxyl groups of each substrate molecule and yields optically active products of dioxidation.

BIOLOGICAL STEREOSPECIFICITIES IN α,ε-DIAMINOPIMELATE METABOLISM: AN EXAMPLE OF STEREOSPECIFICITIES AT A $C_{aa(+b)(-b)}$ *PRO*-CHIRAL CENTER

In Chapter V we discussed the existence of uncommon *pro*-chiral centers which could be designated as $C_{aa(+b)(-b)}$. The known metabolism of the α,ε-diaminopimelates may be used to demonstrate that biological stereospecificity toward such $C_{aa(+b)(-b)}$ *pro*-chiral centers occurs, obeying the same rules as biological differentiation at the more common classes of *pro*-chiral centers. The major biosynthetic route to lysine in bacteria and higher plants proceeds through L,L-α,ε-diaminopimelate and *meso*-α,ε-diaminopimelate, as outlined in Figure VI-4. The α,ε-diaminopimelic acids possess chiral centers at the α and ε carbons, but they only exist in the three distinct stereoisomeric forms illustrated in Figure VI-5. As is apparent from the projected structures in Figure VI-5, the L,L and D,D-α,ε-diaminopimelic acids lack elements of reflective symmetry and are optically active enantiomers; the *meso*-α,ε-diaminopimelic acid, on the other hand, possesses a mirror plane and is an optically inactive diastereomer. The L,L and D,D structures possess C_2 rotational axes that permit the rotational superpositioning of the α and ε chiral centers; whereas the *meso* acid lacks all elements of rotational symmetry.

In considering potential manifestations of biological stereospecificity in

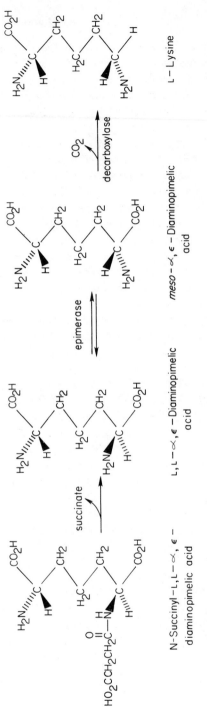

Figure VI-4. Biosynthesis of L-lysine

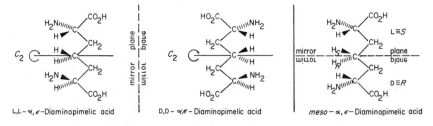

L,L-α,ϵ-Diaminopimelic acid

D,D-α,ϵ-Diaminopimelic acid

meso-α,ϵ-Diaminopimelic acid

Figure VI-5. The α,ϵ-diaminopimelic acid stereoisomers.

the metabolism of these diaminopimelates, it is instructive to analyze the stereochemical relationships at the central γ-methylene position. In each of the three α,ϵ-diaminopimelate stereoisomers, the γ-carbon bears paired methylene hydrogen atoms and paired, *inherently chiral*, amino acid groupings. In both the L,L and D,D-α,ϵ-diaminopimelate structures, Hirschmann's rotational superpositioning test reveals that the paired methylene hydrogens are superimposable and thus equivalent, and that the paired, inherently chiral, amino acid groups are also superimposable and equivalent. Thus the central γ-methylene positions of the L,L and of the D,D-α,ϵ-diaminopimelates may be represented as $C_{aa(+b)(+b)}$ and $C_{aa(-b)(-b)}$. Such non-*pro*-chiral centers have already been mentioned (Chapter V, p. 135). In contrast, *meso*-α,ϵ-diaminopimelate lacks rotational symmetry; therefore, the paired γ-methylene hydrogens and the paired, inherently chiral, amino acid groups of this structure comprise nonequivalent paired functions. The γ-methylene carbon of the *meso* acid is thus a representative of the $C_{aa(+b)(-b)}$ class of *pro*-chiral center previously described (Chapter V, pp. 134–135).

The enzymatic reactions in the lysine biosynthesis route of Figure VI-4 which involve α,ϵ-diaminopimelates have been found to display the biological stereospecificities that are anticipated on the basis of this stereochemical analysis.[8] For example, the diaminopimelate epimerase enzyme (2,6-L,L-Diaminopimelate 2-epimerase), which catalyzes the interconversion of L,L-diaminopimelate and *meso*-diaminopimelate, has a steric specificity for these two diaminopimelate substrates. The D,D-α,ϵ-diaminopimelate stereoisomer is not epimerized by the enzyme, and this acid does not inhibit the epimerase enzyme. Such biological stereospecificity between the enantiomeric L,L and D,D stereoisomers is naturally expected. When the epimerase enzyme acts on the reactive L,L-α,ϵ-diaminopimelate, reversal of the L configuration at either the α or the ϵ chiral center yields the *meso*-α,ϵ-diaminopimelate product (Figure VI-5). Since the α and ϵ centers of L,L-α,ϵ-diaminopimelate are geometrically equivalent, no biological differentiation between them is possible.

Considering the reverse epimerization process, the conversion of *meso*-α,ϵ-diaminopimelate to L,L-α,ϵ-diaminopimelate, the observed biological stereospecificity between the diastereomeric *meso* and D,D stereoisomers is also expected. When the epimerase enzyme acts on the reactive *meso*-α,ϵ-diaminopimelate substrate, however, one *specific chiral center* must be reversed to stereospecifically yield the L,L-α,ϵ-diaminopimelate product. That the γ-carbon atom of the *meso*-α,ϵ-diaminopimelic acid is an example of the unusual but valid $C_{aa(+b)(-b)}$ class of saturated *pro*-chiral center has been emphasized by designating the paired methylene hydrogens at this center [the (a) (a) paired groups] as H_R and H_S in Figure VI-5.* Stereospecific replacement of these paired hydrogens would yield distinct diastereomeric products (cf. Figure V-13). The $(+b)$ and $(-b)$ paired groups of the $C_{aa(+b)(-b)}$ *pro*-chiral center of *meso*-α,ϵ-diaminopimelate are the two alkyl chains bearing the inherently chiral amino acid groupings of opposite absolute configuration (Figure VI-5). These paired groups, $(+b)$ $(-b)$, are not rotationally superimposable but are related geometrically by the molecular mirror plane. In contrast to the equivalent $(+b)$ $(+b)$ and $(-b)$ $(-b)$ paired groups of the L,L and D,D-α,ϵ-diaminopimelates therefore, the amino acid side chains of the *meso*-α,ϵ-diaminopimelate are enantiotopically related paired groups. The stereospecific epimerization of only the D amino acid center in *meso*-α,ϵ-diaminopimelate to produce L,L-α,ϵ-diaminopimelate (Figure VI-4) in the diaminopimelate epimerase reaction is thus consistent with the expected enzymatic differentiation between nonequivalent enantiotopic paired chemical groups.

The decarboxylation of *meso*-α,ϵ-diaminopimelate to yield L-lysine by diaminopimelate decarboxylase (*meso*-2,6-Diaminopimelate carboxy-lyase) as outlined in the biosynthetic scheme of Figure VI-4 displays comparable stereospecificities. The diaminopimelate decarboxylase enzyme has a steric specificity for the *meso*-diaminopimelate diastereomer as a substrate. It should also be apparent from Figures VI-4 and VI-5 that the amino acid grouping of *meso*-α,ϵ-diaminopimelate with the D (R) absolute configuration must be stereospecifically decarboxylated to yield L-lysine. Stereospecific decarboxylation of the amino acid grouping with the L absolute configuration would yield D-lysine, whereas random decarboxylation would yield a racemic lysine product. Diaminopimelate decarboxylase, therefore, must also differentiate between the enantiotopic paired amino acid groupings of the γ-*pro*-chiral center of *meso*-α,ϵ-diaminopimelate.

The enzymatic differentiation between the enantiotopic paired groups of the $C_{aa(+b)(-b)}$ *pro*-chiral center of *meso*-α,ϵ-diaminopimelate displayed in

---

* When applying the Cahn-Ingold-Prelog system of configurational nomenclature to such molecules, a rule that R chirality precedes S chirality governs the ordering of groups.

the diaminopimelate epimerase and diaminopimelate decarboxylase reactions leads to the productions of particular stereoisomeric forms of the products. This observation serves to emphasize again the close relationship that exists between the two aspects of biological stereospecificity—production or conversion of one stereoisomeric molecular structure, and differentiation between the chemically like, paired substituents of a substrate molecule.

BIOLOGICAL STEREOSPECIFIC REACTIONS OF (R) AND (S)-ACETYL CoA SUBSTRATES

The recent investigations of enzymatic reactions of acetate substrates possessing chiral methyl groups have led to truly significant advances in the field of biological stereospecificity. For example, the citrate synthetase reaction is central to the operation of the TCA cycle. This reaction converts the methyl group of acetyl CoA into a methylene group ($-CH_2-$) of the resulting citric acid product. Such a methyl–methylene interconversion may be represented schematically by the process of Figure VI-6. In addition to the TCA cycle stereochemistry summarized in Figure VI-1, we can ask if, in the citrate synthetase reaction, there is a demonstrable stereospecific relationship between the original carbon–hydrogen bond of the acetyl CoA methyl group and the new carbon–carbon bond in the citrate product. If the new carbon–carbon bond of the methylene group possesses the same relative stereochemical orientation as the carbon–hydrogen bond it replaced, then we would define the process as involving *retention* of configuration at the condensing methyl carbon atom. If the new carbon–carbon bond possesses the opposite stereochemical orientation as the carbon–hydrogen bond it replaced, we would define the condensation process as involving *inversion* of the methyl carbon atom configuration. Clearly these definitions of retention and inversion could also be applied to the reverse methylene to methyl conversions, if the configuration of the new carbon–hydrogen bond could be determined and compared to the stereochemistry of the carbon–carbon bond cleaved in the reverse (lyase) process. To be sure, the three carbon–hydrogen bonds of acetate molecules possessing unsubstituted methyl groups may be super-

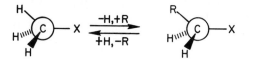

Figure VI-6. Methyl–methylene interconversion typified by the citrate synthetase reaction.

imposed by free rotation and are thus stereochemically equivalent. In such cases biological stereospecificity of citrate synthetase toward the acetyl CoA methyl group, if it were to exist, could not be detected. An acetyl CoA substrate possessing a substituted methyl group, however, can have either a *pro*-chiral or a chiral center at the methyl carbon. In such substrates it is possible to determine a stereochemical relationship between the original methyl group and the methylene group of the product. For example, in the discussion of monofluorocitrates in Chapter IV it was noted that the specific biological formation of $(2R,3R)$-2-fluorocitrate from monofluoroacetyl CoA established such a stereochemical relationship. This result indicated that either the *pro-R* hydrogen was replaced with retention of configuration or the diastereotopic *pro-S* hydrogen was replaced with inversion of configuration.

Two groups of workers have recently used acetic acid substrates possessing chiral methyl groups [(R) and (S)-2-deuterio-2-tritioacetates] to establish that the citrate synthetase reaction involves an inversion of configuration at the condensing methyl group.[9-12] This noteworthy achievement is an indication of the great advances that have been made in the study of biological stereospecificity since Ogston published his classical analysis in 1948.

The reactions involving acetates with asymmetric methyl groups investigated to date are those catalyzed by (si)-citrate synthetase,[11,12] (re)-citrate synthetase,[11] citrate lyase,[11] ATP citrate lyase,[11] and malate synthetase.[9,10] It is perhaps of mechanistic significance that all the acetyl CoA condensations or acetate-forming lyase reactions investigated thus far have been found to result in inversion at the methyl group (synthetases) or at the methylene group (lyases) undergoing reaction.

In addition to the mechanistically important stereochemical conclusions of such studies, investigations of this type always involve several other interesting features. These include the method of preparing and analyzing the appropriate substrate, the method of detecting a stereospecific biological reaction, and the method of determining the absolute stereochemistry of the product. Two different synthetic approaches were used to prepare the chiral 2-deuterio-2-tritioacetate substrates needed for the investigations cited. One synthetic approach to the two chiral acetates was based on biological stereospecific reactions;[10,12] the second approach, utilized in two related syntheses,[9,11] was based entirely on stereospecific chemical procedures. The route to chiral acetate molecules, which involved biological stereospecific reactions, is outlined in Figure VI-7.

Lactate dehydrogenase catalyzes the addition of hydride to the *re* face of the aldehydic carbon of glyoxylate. Figure VI-7 illustrates how this stereospecific biological reduction, when carried out with 2-tritioglyoxylate, will yield (S)-2-tritioglycolate. The glycolate methyl ester was then reduced to ethylene glycol with $LiAlH_4$ and the monobrosylate ester prepared. Owing

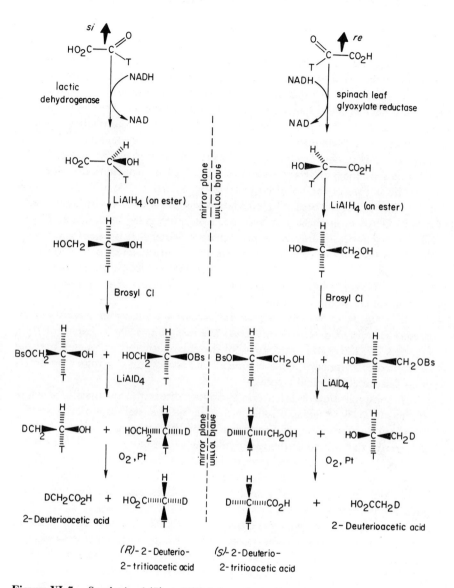

Figure VI-7. Synthesis of (R) and (S)-2-deuterio-2-tritioacetates via biological stereo-specific reactions.[10]

to the tritium labeling originally present, this series of reactions produced a mixture of monobrosylate esters—(1S)-(1-tritio-2-hydroxy)ethyl brosylate and (2S)-(2-tritio-2-hydroxy)ethyl brosylate. The mixture of monobrosylate esters was then treated with LiAlD$_4$ in tetrahydrofuran to reductively cleave the ester linkage. Since this process is known to proceed with inversion at the reacting center, the ethanol produced was a mixture of 2-deuterio-1-tritio-ethanol and (R)-2-deuterio-2-tritioethanol. Careful oxidation then yielded the desired (R)-2-deuterio-2-tritioacetate and 2-deuterioacetate. The fact that a mixture of chiral and *pro*-chiral acetate molecules was produced by the outlined sequence, although potentially troublesome conceptually, does not alter the nature of the required experimental analysis. As we discuss later, only a fraction of the original glyoxylate molecules were actually labeled with tritium. Thus even the acetic acid product represented as chiral in Figure VI-7 will consist mainly of nontritiated, 2-deuterioacetic acid molecules. Since the analyses of reactions involving the chiral acetate substrates must already have accounted for this, the additional source of nontritiated, *pro*-chiral, acetic acid molecules inherent to the synthetic scheme of Figure VI-7 does not affect the final results.

   It is significant, however, that all the 2-deuterio-2-tritioacetate produced by the sequence just described possesses the R absolute configuration. As is apparent from Figure VI-7, the *re* stereospecificity of the enzymatic reduction step and the stereospecific inversion of the chemical monobrosylate cleavage step determine this stereochemical result.

   The preparation of 2-deuterio-2-tritioacetate with the opposite absolute configuration was accomplished by *reversing* the stereospecificity of the enzymatic reduction step. Spinach leaf gloxylate reductase catalyzes hydride addition to the *si* face of the aldehydic carbon of glyoxylate. Reduction of 2-tritioglyoxylate with this enzymatic system therefore produced (R)-2-tritioglycolate. Modification of this intermediate by the chemical steps previously described produced a mixture of (S)-2-deuterio-2-tritioacetate and 2-deuterioacetate.

   The second group of investigators studying the biological reaction stereospecificities with chiral acetate substrates prepared the necessary acetate substrates by using only stereospecific chemical processes.[9,11] The most recent of their two published procedures is outlined in Figure VI-8.

   In this stereospecific synthesis *trans*-2-bromostyrene was converted into *trans*-2-tritio-1-phenylethylene by T$_2$O quenching of the lithium derivative. Epoxidation of the phenylethylene (a stereospecific *syn* reaction) and reductive opening of the epoxide with LiAlD$_4$ (a stereospecific reaction involving inversion) led to a mixture of the (1S,2S) and (1R,2R)-2-deuterio-2-tritio-1-phenylethanol enantiomers as illustrated in Figure VI-8. This mixture then was resolved into its separate (1S,2S) and (1R,2R) components by fractional

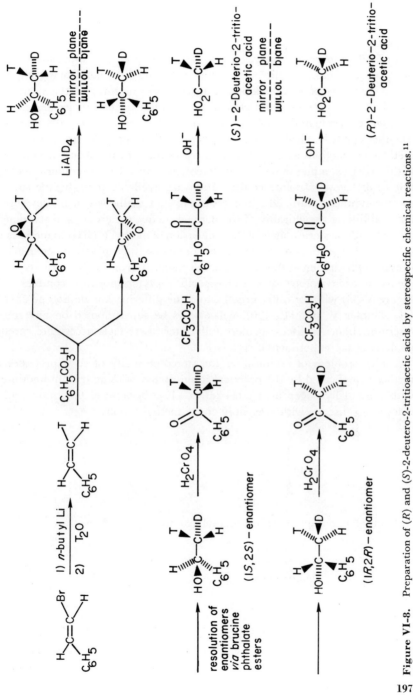

Figure VI-8. Preparation of (R) and (S)-2-deutero-2-tritioacetic acids by stereospecific chemical reactions.[11]

197

crystallization of the diastereomeric brucine phthalate esters. Oxidation of the separated 1-phenylethanol enantiomers produced the phenyl (R) and (S)-acetates, which, upon hydrolysis, yielded the desired (R) and (S)-2-deuterio-2-tritioacetate molecules. Again, it should be emphasized that, owing to the impracticality of working with 100% tritiated materials, the actual products of the reaction sequence were mixtures of the indicated chiral acetate molecules and *pro*-chiral, 2-deuterioacetate molecules.

The experimental problems that have to be overcome in order to establish the stereochemical course of reactions that interconvert methyl and methylene functions are significantly greater than the problems encountered with earlier examples of biological stereospecificity. Even after chiral methyl groups of known absolute configuration are available (e.g., by the stereospecific syntheses outlined), and assuming that methods can be developed for establishing the chirality of the derived methylene group, still the determination of the steric course of the conversion of methyl $(-CH_3)$ to methylene $(-CH_2-)$ is dependent on an appreciable reaction discrimination between the isotopes of hydrogen at the chiral methyl center. Certainly the steric requirements of each isotope in a stereospecific enzyme-substrate complex (cf. Figure V-35) will not differ greatly, and the differing kinetic isotope effects (cf. Chapter V, pp. 118–120), will always be superimposed on any steric discrimination. As discussed more fully later, these facts necessitate careful analysis of the experimental results.

The problem of establishing the stereospecificity of such processes is further complicated by the practical impossibility of using tritium undiluted with normal hydrogen as a label (cf. p. 117). Because of this, most molecules of a substrate with ostensibly chiral methyl groups

will actually contain *pro*-chiral methyl groups

in which normal hydrogen replaces the tritium label. Thus optical rotation methods are ruled out as a means of establishing or analyzing the chiralities. The difficulty can be overcome by ensuring that every methyl group containing a tritium atom also contains a deuterium label and then by relating the determination of chirality to radioactive measurements. In this way only those molecules which actually contain tritium and therefore possess truly chiral methyl groups will be analyzed. *If* the reaction pathway *then* involves an appreciable isotopic discrimination between hydrogen and deuterium, the radioactive measurements can be used to establish the stereochemical course. The schematic representation in Figure VI-9 will help to clarify these points.

Figure VI-9 illustrates possible stereochemical relationships for a reaction that converts a chiral acetate methyl group into a methylene function with overall *retention of configuration* at the methyl carbon atom. Conformations *1–6* represent possible reactive orientations of the chiral acetate substrates and the enzyme. If there is no steric differentiation between ^1_1H, ^2_1H, and ^3_1H in the enzyme-substrate complex, then the reactive conformations *1–6* will be equally probable. If, in addition, the rates of product formation from all conformations are the same (all k's are equal), then the stereospecific course of the reaction (retention) will not be detected even if acetates substituted with chiral methyl groups are used. Consider, for example, only the products

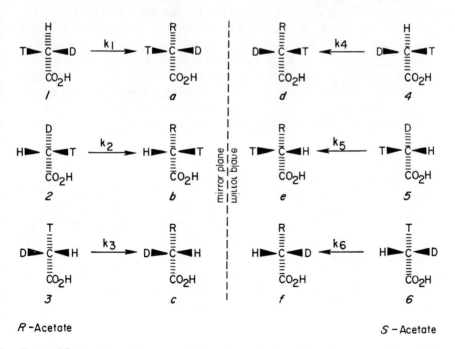

R –Acetate   S – Acetate

Figure VI-9. Schematic representation of a methyl–methylene interconversion of (R) and (S)-acetate substrates with retention of configuration.

bearing a tritium-substituted methylene group: a, b, d, and e. Assuming that the substituent group "R" is lower in priority then a carboxyl group, then product a, formed at rate k_1, bears an R tritium label on the resulting methylene. Product b, however, is derived from the same original (R)-acetate at rate k_2 and bears an S tritium label on the resulting methylene. If, as proposed previously, the 1 and 2 conformations are of equal energies and thus of equal populations *and* $k_1 = k_2$, then clearly the (R)-2-deuterio-2-tritioacetate substrate will lead to products whose methylene groups will be substituted with equal amounts of R and S tritium label. Exactly the same situation will result from reaction of the enantiomeric (S)-2-deuterio-2-tritioacetate substrate. It should be readily appreciated that such a result—with no discernible stereochemical difference in the reaction of (R) or (S)-acetates—could never distinguish between a stereospecific process of inversion, a stereospecific process of retention, or a totally random, nonstereospecific process.

Consider, however, the experimental situation if conformation populations are of equal energy but the normal primary kinetic isotope effect prevails; that is, $k_H > k_D > k_T$. Now, despite reactive conformations 1 and 2 being present in equal amounts, $k_1 > k_2$ (breaking a carbon–hydrogen bond being easier than breaking a carbon–deuterium bond; cf. Chapter V, pp. 118–119) and the (R)-acetate substrate will lead to a product with most of the tritium on the resulting methylene possessing the R configuration ($a > b$). At the same time, since $k_4 > k_5$, the (S)-acetate substrate will lead to a product with most of the tritium in the product possessing the S configuration ($d > e$). In this way, *when an appreciable isotopic discrimination is associated with the methyl–methylene interconversion being investigated, the different stereochemical results obtained from R and S chiral methyl substrates can be used to establish the stereospecificity of the process.*

The foregoing illustration emphasizes the dependence of this experimental approach on an appreciable isotopic discrimination during the reaction. It should also be emphasized that because the overall observed isotopic discrimination is composed of several variable factors, the interpretations of differentiable methylene labeling in such experiments are not entirely straightforward. For example, if the reaction being investigated displayed an inverse kinetic isotope effect* ($k_D > k_H$), then k_2 would be greater than k_1. Unless the interpretation took this into account, the observed experimental results would lead to an incorrect stereochemical conclusion. Alternatively, if in the previous illustration, for unknown steric reasons, reactive conformation 1 was preferred over conformation 2 (and 4 over 5),

---

* Unusual, but not unknown, cf. W. P. Jencks, in *Catalysis in Chemistry and Enzymology*, pp. 264–267, McGraw-Hill, 1969.

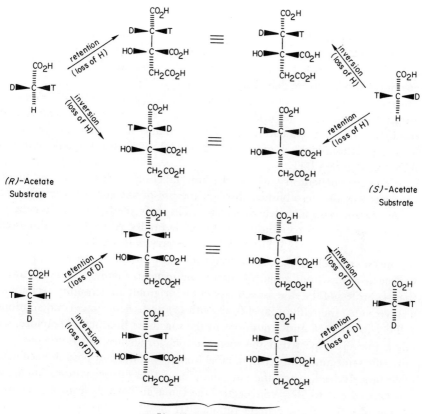

possible tritiocitric acid products

Figure VI-10. Possible results of stereospecific condensations of R and S chiral acetates by si-citrate synthetase.

then again the stereochemical results *could* be the opposite—although the reaction would still display a normal kinetic isotope effect and proceed with retention of configuration.

Returning now to the si or (S)-citrate synthetase reaction, the potential results of stereospecific condensations involving the R and S chiral acetate substrates are summarized in Figure VI-10. For reasons given earlier, Figure VI-10 only summarizes the possible stereochemical results for product molecules that retain a tritium label. Establishing the stereochemical course of the citrate synthetase reaction will require that the experimental results from the R and S acetate substrates be significantly different and that the observed differences be interpreted in terms of the stereochemical predictions of Figure VI-10.

In their stereochemical analysis of the tritium-labeled citrate molecules, Rétey et al.[12] made use of the biological stereospecificities of additional TCA enzymes. The labeled citrate was first converted to α-ketoglutarate by aconitase and isocitrate dehydrogenase catalysis. As outlined in Figure VI-1, this sequence leaves the methylene hydrogen atoms on the *pro-S* branch of citrate undisturbed. The α-ketoglutarate was then chemically oxidized to succinate. The various tritium labeled succinates that could be produced from the tritium-labeled citrates of Figure VI-10 by this sequence are depicted in Figure VI-11.

The stereochemistry of the tritium labeling within these succinate products was then established through the use of yet another stereospecific TCA enzyme, succinic dehydrogenase. The four methylene hydrogens of succinate occur as two enantiotopically related pairs. As previously discussed (Chapter V, pp. 131–133; Figure VI-1), the enzymatic oxidation of succinate by succinate dehydrogenase stereospecifically removes *pro-R* and *pro-S* hydrogen atoms from adjacent methylene groups. Figure VI-11 illustrates that regardless of the stereospecificity of the original condensation (retention or inversion), the tritium label from both (S) and (R)-acetate condensations will occupy both R and S positions in the succinate products. Therefore, we might initially expect that the succinate dehydrogenase reaction would remove the same amount of tritium label from any of the four possible succinate molecules. Indeed, when the enzymatic oxidation of succinate to fumarate was carried out to 70% conversion, the fumarate formed from a reference sample of (R)-2-tritiosuccinate retained $52 \pm 2\%$ of the original tritium, whereas the fumarate formed from a reference sample of (S)-2-tritiosuccinic acid retained $56 \pm 2\%$ of the original label. These results are due to compensating effects. Since the dehydrogenase reaction simultaneously removes a *pro-R* and a *pro-S* hydrogen, tritium will be removed from both (R) and (S)-2-tritiosuccinates. However, owing to the kinetic isotope effect $k_H > k_T$, and therefore *intramolecular* competition will favor loss of hydrogen and hydrogen rather than hydrogen and tritium. This effect will tend to *enrich* the amount of tritium remaining in the fumarate products. At the same time, only a small fraction of the labeled molecules actually bear a tritium atom (see above), and *intermolecular competition* between the labeled molecules and the larger pool of more rapidly reacting unlabeled molecules will tend to *delay* the appearance of tritium in the product. That the kinetic isotope effect favors overall retention of tritium in the fumarate product was shown by the $58 \pm 2\%$ retention value observed from a racemic (RS)-2-tritiosuccinate reference substrate.

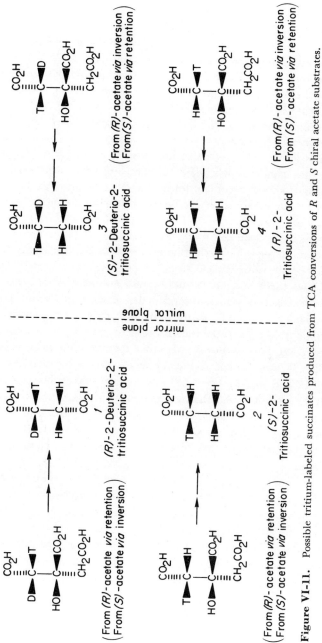

Figure VI-11. Possible tritium-labeled succinates produced from TCA conversions of *R* and *S* chiral acetate substrates.

203

In contrast to the quite similar tritium retention values observed from the (R), (S), and (RS)-2-tritiosuccinate reference samples, the fumarate produced in the (R) chiral acetate experiment showed a $31 \pm 1\%$ retention of the original succinate tritium; the fumarate produced in the (S) chiral acetate experiment, on the other hand, showed $49 \pm 2\%$ tritium retention. (Again, the enzymatic oxidations were carried out to 70% conversion.) These contrasting results indicate differing stereospecific tritium labeling in the two samples. The problem is to interpret these differing results in terms of the stereochemical possibilities of Figure VI-11.

The results obtained with the reference compounds make it apparent that neither form 2 [(S)-2-tritiosuccinate] nor form 4 [(R)-2-tritiosuccinate] could account for the 31% retention observed in the fumarate derived from the (R)-acetate condensation. The amounts of tritium retained in the fumarate products from succinate forms 1 and 3 of Figure VI-11, however, will depend on the *additional kinetic isotope effects generated by the presence of a deuterium atom on the methylene position bearing the tritium.* The overall retention of tritium will depend on the ratio of the isotope effect for removal of the deuterium atom to the isotope effect for removal of a tritium atom of opposite configuration on the same carbon atom (i.e., $k_{HR}/k_{DR} : k_{HS}/k_{TS}$). It has been established that there is a substantial kinetic isotope effect in the succinic dehydrogenase reaction which discriminates against deuterium in the R configuration ($k_{HR}/k_{DR} = 5.3$). At the same time, there is only a small isotope effect discriminating against the removal of deuterium in the S configuration ($k_{HS}/k_{DS} = 1.35$).[13] By analogy with other established isotope effects, the corresponding *tritium* kinetic isotope effects can be anticipated to be $k_{HR}/k_{TR} \simeq 12$ and $k_{HS}/k_{TS} \simeq 1.5$.[14] During intramolecular competitions, therefore, there will be an enhanced tendency to lose an S tritium from methylene positions also bearing an R deuterium [i.e., structure 3 of Figure VI-11, (S)-2-deuterio-2-tritiosuccinate, in which $k_{HR}/k_{DR} : k_{HS}/k_{TS} \simeq 5.3 : 1.5^*$]. In contrast, since the kinetic isotope effect for a deuterium in the S configuration is small ($k_{HS}/k_{DS} = 1.35$), structure 1 of Figure VI-11 (in which $k_{HS}/k_{DS} : k_{HR}/k_{TR} \simeq 1.35 : 12$) will give results that differ only slightly from those of the nondeuterated (R)-2-tritiosuccinate (structure 4). The extensive loss of tritium from the succinate sample derived from (R) chiral acetate (69%), therefore, *uniquely* establishes this material as containing appreciable amounts of the (S)-2-deuterio-2-tritiosuccinate (structure 3). As indicated in Figures VI-10 and VI-11, these results require that the *si*-citrate synthetase reaction occur with inversion at the condensing methyl carbon atom.

It was independently established by Eggerer et al.[11] that the *si*-citrate

---

* Remember that the *larger* the isotope effect, the *more difficult* it is to break the bond to the isotopically substituted position.

synthetase reactions occur with inversion at the methyl carbon. This group determined the stereochemistry of the tritium-labeled citrates represented in Figure VI-10 by a totally different route. In the initial step of the stereochemical analysis, the citrate samples were treated with citrate lyase enzyme. This enzyme catalyzes the cleavage of citrate into acetate and oxaloacetate. The biological stereospecificity of this lyase is such that the *pro-S* carboxymethyl branch is selectively converted into the acetate (Chapter V, p. 157). Since this *pro-S* branch is originally derived from the acetyl CoA substrate, a stereospecific citrate lyase reaction in H_2O will produce chiral acetate samples from the 2-deuterio-2-tritiocitrate samples of Figure VI-10. Thus, we see that the sequence of a citrate synthetase reaction followed by a citrate lyase reaction condenses and then regenerates the chiral acetates. However, it was impossible to establish the chirality of the acetates regenerated by this sequence by optical methods (see earlier discussion). Therefore, the acetates from the lyase reaction were condensed with glyoxylate in the presence of malate synthetase to form malate. Finally, the tritium-labeled malate samples were converted to fumarate by fumarase. The amount of tritium lost during the final malate–fumarate conversion, and a knowledge of the biological stereospecificities of all the enzymes utilized in this analytical sequence—citrate lyase, malate synthetase, and fumarase—permitted the workers to assign the stereochemistry of the original tritiated citrate samples. This overall analytical sequence is summarized in Figure VI-12.

The malate synthetase reaction was actually the first reaction investigated with (R) and (S)-2-deuterio-2-tritioacetate substrates.[9,10] Prior to the experiments involving chiral acetates, it was known that the malate synthetase reaction involves condensation on the *si* face of the glyoxylate aldehydic carbon to form L-malate [(S)-malate]. It was also known that the fumarase catalyzes the *anti* hydration-dehydration of fumarate and L-malate, respectively (cf., pp. 123–126). In the reversible hydration-dehydration of L-malate by fumarase, therefore, the *pro-R* hydrogen of the malate substrate will be stereospecifically removed or exchanged. The malate synthetase and fumarase reactions can thus be represented as in Figure VI-13.

Figure VI-13 illustrates the expected behavior in the malate synthetase reaction *if* (S)-2-deuterio-2-tritioacetyl CoA condenses stereospecifically (either with inversion *a*, or retention *b*, at the methyl carbon), *and if* this reaction involves appreciable discrimination among the various isotopes of hydrogen due to a normal kinetic isotope effect ($k_1 > k_2$; $k_1' > k_2'$). Under such conditions, more tritium derived from the (S)-acetate substrate will be situated in the fumarase labile R configuration in malate when the malate synthetase reaction involves inversion (pathway *a*). The relative amount of tritium lost during the reversible fumarase-catalyzed malate to fumarate interconversions in H_2O were determined by adding a reference amount of acetate-1-^{14}C to the original reaction mixture. The ^{14}C/^{3}H ratios determined

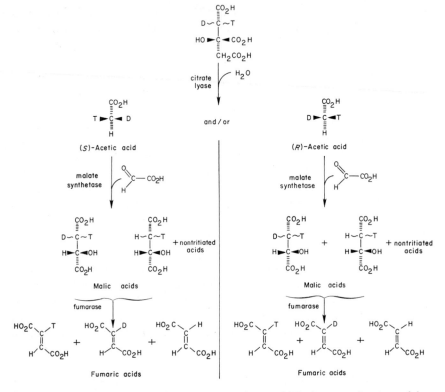

Figure VI-12. Analytical reaction sequence used to establish the stereochemistry of the tritiated citrate samples from chiral acetate condensations.[11]

in the malate substrate and in the malate and fumarate products were used to calculate the percentage of malate tritium label removed and exchanged by fumarase. Using this procedure, it was established that $76.5 \pm 0.6\%$ of the malate tritium label was lost in the fumarase reaction in H_2O when (S)-acetate was the original precursor. In the case of malate labeled with tritium from (R)-acetate, however, only $25.3 \pm 0.8\%$ of the tritium was lost when the fumarase-catalyzed interconversions were carried out in H_2O.*

---

* Using the (R) and (S)-2-deuterio-2-tritioacetate substrates synthesized with the aid of a biological (enzymatic) stereospecific process, Lüthy et al. found that about 90% of the malate tritium label from (S)-acetate and only about 10% of the malate label from (R)-acetate was lost during the fumarase reaction.[10] In general, biological processes manifest greater stereospecificity than do strictly chemical reactions, and perhaps the greater differentiation between the products in this case reflects a somewhat greater stereochemical purity of the chiral acetate substrates produced by the biochemically based synthesis (cf., however, Ref. 15, p. 12).

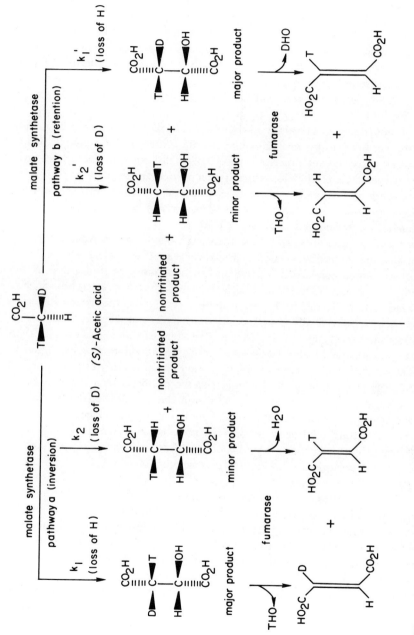

Figure VI-13. Potential stereospecific malate synthetase and fumarase reactions of (*S*)-2-deuterio-2-tritioacetate.

207

Figure VI-13 illustrates that the extensive loss of tritium from malate synthesized from the (*S*)-acetate substrate is consistent with malate synthesis involving an inversion at the condensing methyl carbon atom. As discussed previously with regard to the potential processes outlined in Figure VI-9, it is the observed *differences* in the results from the *R* and *S* chiral acetates that permit valid stereochemical conclusions. The reader should therefore construct a diagram analogous to Figure VI-13 for the (*R*)-acetate substrate, to ascertain that the high degree of retention of the tritium label observed in this malate product is also consistent with a malate synthesis with stereospecific inversion.

Once it has been established that malate formation catalyzed by malate synthetase involves an inversion, the malate synthetase–fumarase sequence becomes a convenient procedure to establish the absolute configuration in acetate molecules of unknown chirality. In particular, the malate synthetase–fumarase sequence can be utilized to establish the configuration of the acetate samples produced by a citrate lyase reaction in H_2O upon the 2-deuterio-2-tritiocitrates represented in Figure VI-10.

Condensation processes that convert methyl groups into methylene functions are, of course, related to the reverse process, a cleavage (or lyase) reaction, which converts a methylene function into a methyl group. If the methylene group affected in a lyase reaction has been stereospecifically labeled with differing hydrogen isotopes, and if the methylene cleavage process stereospecifically introduces a third hydrogen isotope, then the resulting methyl group will be chiral. The stereochemical analyses of such lyase processes at labeled methylene groups are therefore closely related to the stereochemical analyses of chiral methyl group condensations described previously. Establishing the biological stereospecificity of a lyase reaction, however, depends *only* on determining both the chirality of the methylene group affected and the chirality of the methyl group produced by introduction of the third hydrogen isotope. In contrast to the methyl–methylene interconversions previously noted, isotopic discrimination is *not* required in order to establish the steric course of a lyase reaction that involves a methylene–methyl interconversion.

Because the malate synthetase–fumarase sequence can now be used to determine the configuration of acetate molecules with chiral methyl groups, establishing the stereospecificity of the citrate lyase reaction is only dependent on producing a citrate substrate labeled with hydrogen isotopes of known configuration at the methylene carbon of the *pro-S* carboxymethyl branch. Eggerer et al. prepared the necessary citrate substrates by condensing stereospecifically labeled oxaloacetate molecules with acetyl CoA in the presence of the uncommon (*R*) or *re*-citrate synthetase (cf. Table V-1). This citrate synthetase stereospecifically forms the *pro-S* branch of citrate from oxaloacetate. Therefore, the stereospecifically labeled methylene of oxaloacetate becomes that methylene of citrate (*pro-S*) specifically converted to methyl by citrate lyase. The overall sequence utilized is depicted in Figure VI-14.

(2S,3R)-3-Tritiomalate and (2S,3S)-2,3-ditritiomalate samples were prepared with fumarase, by treating fumarate in tritiated water and by treating 2,3-ditritiofumarate in water (*anti*-hydration reaction, cf. Figure VI-1). These specifically labeled malate substrates were then treated with malate dehydrogenase and NAD in the presence of acetyl CoA and *re*-citrate synthetase. In this way the labeled oxaloacetate produced in the dehydrogenase reaction was rapidly and irreversibly converted into citrate. As illustrated, this sequence produced citrates stereospecifically tritiated at the *pro-S* methylene group. Each of the labeled citrate samples was then cleaved with citrate lyase in D_2O to produce oxaloacetate and chiral acetates. The chiralities of the acetates produced by each reaction sequence were then determined by the malate synthetase–fumarase assay system just described. Beginning with the (2S,3R)-3-tritiomalate, chiral acetate was produced which led to a retention of 60.7 ± 1.4% of the tritium label in malate-fumarate after the malate synthetase–fumarase assay. The absolute configuration of the chiral acetate in this sample was therefore predominantly R (see above discussion). From (2S,3S)-2,3-ditritiomalate, chiral acetate was produced which led to a retention of 43.6 ± 0.7% of the tritium in malate-fumarate after the malate synthetase–fumarase assay. The chiral acetate in this sample was therefore predominantly S. As illustrated in Figure VI-14, these results require that the citrate lyase reaction proceed with an inversion of configuration at the methylene carbon undergoing conversion to methyl.

The knowledge of the biological stereospecificities of the citrate lyase, malate synthetase, and fumarase reactions developed previously *now* permits us to use this sequence of enzymatic reactions to analyze the steric course of the *si*-citrate synthetase reaction.

The (S) or *si*-citrate synthetase was used to form citrate from (R)-3-deuterio-2-tritioacetate. The citrate produced was cleaved with citrate lyase in the presence of H_2O and the chiral acetate released analyzed via the malate synthetase–fumarase sequence. The malate-fumarate produced upon incubation with fumarase in H_2O retained 65.0 ± 0.7% of the tritium label originally present in the malate product. This result established that (R)-acetate was regenerated by the lyase reaction upon citrate biosynthesized from (R)-acetate. When (S)-2-deuterio-2-tritioacetate was utilized with the *si*-citrate synthetase, the malate-fumarate generated by the analysis scheme retained 40.6 ± 0.6% of the original malate tritium label. Thus it could be inferred that, when (S)-acetate was used as the original substrate for the *si*-synthetase enzyme, (S)-acetate was regenerated by the lyase reaction. Since the lyase cleavage involved an inversion, the foregoing sequences—

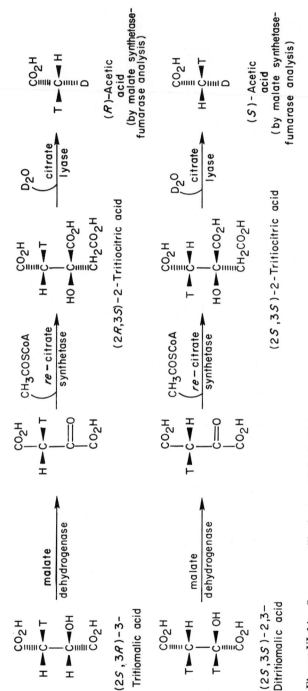

Figure VI-14. Sequence utilized to establish the biological stereospecificity of citrate lyase.

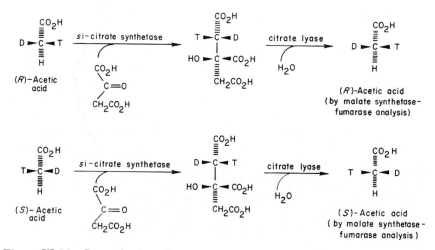

Figure VI-15. Determination of stereospecificity of *si*-citrate synthetase reaction via citrate lyase degradation.

(R)-acetate $\xrightarrow{\text{synthetase}}$ citrate $\xrightarrow{\text{lyase}}$ (R)-acetate and (S)-acetate $\xrightarrow{\text{synthetase}}$ citrate $\xrightarrow{\text{lyase}}$ (S)-acetate—required that the *si*-citrate synthetase reaction also proceed with inversion. (This is schematized in Figure VI-15.) Therefore the results of the independent analysis of *si*-citrate stereospecificity by Eggerer et al. agree with the results reported by Rétey et al. and described earlier in this chapter. It is of mechanistic significance that citrate syntheses proceed with inversion at the condensing methyl group.* It had been previously suggested that the citrate synthetase reaction involved an enolization of the acetyl CoA catalyzed by one of the carboxylate groups of the oxaloacetate.[16,17] Since such a mechanism would lead to a net retention of the configuration at the condensing methyl carbon, this mechanistic proposal is no longer tenable.

BIOLOGICAL STEREOSPECIFICITIES IN THE SQUALENE BIOSYNTHETIC PATHWAYS: RESOLUTION OF 13 OF THE 14 POINTS OF AMBIGUITY

Determination of the biological stereospecificities manifest during the biosynthesis of squalene from mevalonate is one of the classics of stereochemical analysis. This sequence involves only seven different enzymatic reactions, and only the initial intermediates in the pathway are dissymmetric;

---

* An inversion pathway has also been indicated for the unusual *re* or (R)-citrate synthetase.[11]

nevertheless, 14 separate stages may be identified in this biosynthetic sequence where biological stereospecificity is to be expected. The absolute stereospecificity at all but one of these 14 stages has been determined. Although most investigations of biological stereospecificity have been concerned with individual enzymatic reactions, our knowledge of the stereospecificities of the entire mevalonate–squalene pathway is almost totally due to one coordinated research effort directed by Popják and Cornforth. In the first Ciba Medal Lecture these scientists outlined the elegant experiments they carried out to establish the stereospecificity at 12 of the 14 points of "stereochemical ambiguity" in the squalene biosynthetic pathway.[18] These 14 stages of the mevalonate–squalene pathway where biological stereospecificity is to be expected are numbered in Figures VI-16 and VI-17.

In their analysis of this sequence of biosynthetic reactions, Popják and Cornforth asked the following questions:

Stage 1.   During the conversion of mevalonate 5-pyrophosphate to isopentenyl pyrophosphate, are the hydroxyl and carboxyl groups eliminated in a *syn* or in an *anti* manner?

Stage 2.   When the isopentenyl phrophosphate is isomerized to 3,3-dimethylallyl pyrophosphate, is a hydrogen atom added to the methylene carbon on the *re,re* face or the *si,si* face of the *pro*-chiral double bond?

Stage 3.   In this isopentenyl pyrophosphate to 3,3-dimethylallyl pyrophosphate isomerization, is the H_S or the H_R paired hydrogen atom eliminated from the *pro*-chiral carbon?

Stage 4.   After this isomerization, is the methyl group formed from hydrogen addition to the unsaturated methylene group located *cis* or *trans* to the carbon chain?

Stage 5.   When 3,3-dimethylallyl pyrophosphate condenses with isopententyl pyrophosphate, is the dimethylallyl group added to the *re,re* face or the *si,si* face of the *pro*-chiral double bond of the isopentenyl molecule?

Stage 6.   During this condensation, is an H_R or H_S paired hydrogen atom eliminated?

Stage 7.   Is there retention or inversion of configuration at the C-1 position of the dimethylallyl pryophosphate during the formation of the new carbon–carbon bond?

Stages 8, 9, 10.   Questions analogous to those asked at stages 5, 6, and 7 may be asked with respect to the condensation of geranyl pyrophosphate with isopentenyl pyrophosphate.

Stage 11.   In the reductive coupling of two molecules of farnesyl pyrophosphate to form squalene, is the single hydrogen atom known to be lost from the C-1 position of one of the farnesyl moieties[19] an H_S or an H_R paired atom?

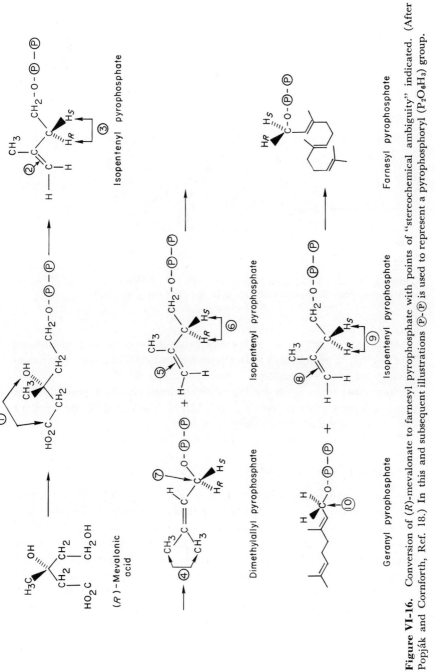

Figure VI-16. Conversion of (*R*)-mevalonate to farnesyl pyrophosphate with points of "stereochemical ambiguity" indicated. (After Popják and Cornforth, Ref. 18.) In this and subsequent illustrations ℗–℗ is used to represent a pyrophosphoryl (P₂O₆H₃) group.

213

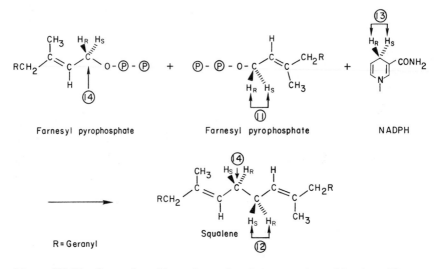

Figure VI-17.   Conversion of farnesyl pyrophosphate to squalene with points of "stereochemical ambiguity" indicated. (After Popják and Cornforth, Ref. 18.)

Stage 12.   In the squalene molecule, what is the absolute configuration of the single hydrogen known to be derived from NADPH during the reductive coupling of two farnesyl pyrophosphate molecules?[19]

Stage 13.   Is the H_S or the H_R paired hydrogen at the 4' position of the reduced dihydronicotinamide ring of NADPH transferred to the squalene molecule in this reductive coupling?

Stage 14.   Is the configuration of the C-1 position of that farnesyl moiety which does not exchange hydrogen in the coupling process inverted or retained when the new carbon–carbon bond is formed?

As indicated, this analysis identifies 14 separate stages of "stereochemical ambiguity" in the squalene biosynthetic pathway. Prior to the investigations of Popják and Cornforth it had been determined only that the methyl group formed in the isopentenyl pyrophosphate to dimethylallyl pyrophosphate conversion is *trans* to the carbon chain (stage 4).[20,21] Popják and Cornforth first established that squalene synthetase is a "class B enzyme" and that the H_S paired hydrogen at the 4' position of the reduced nicotinamide ring of NADPH is transferred to the squalene molecule during the reductive coupling reaction (stage 13).* The preparation of a dissymmetric succinate standard

---

* A more detailed discussion of the biological stereospecificity toward these diastereotopic paired hydrogens of NADPH appears on pages 240–244.

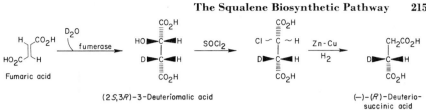

Figure VI-18. Preparation of $(-)$-(R)-2-deuteriosuccinic acid.

of known absolute configuration was vital to this aspect of the investigation. Popják and Cornforth utilized this same dissymmetric succinate standard to resolve the stereochemical ambiguities at stages 7, 10, 12, and 14 of the biosynthetic pathway. The necessary succinate standard was stereospecifically prepared from fumarate by the combination of enzymatic and chemical reactions outlined in Figure VI-18. When fumarate was hydrated with fumarase in the presence of deuterated water, the established biological stereospecificity led to the production of $(2S,3R)$-3-deuteriomalate (cf. Figure VI-1). This enzymatically deuterated malate was esterified and converted into 2-chloro-3-deuteriosuccinate with thionyl chloride, and then the chlorine atom replaced with hydrogen in a reduction catalyzed by zinc-copper couple. Because this chemical sequence does not affect the absolute configuration of the deuterated methylene group, the final succinate product must have been the (R)-2-deuteriosuccinate pictured in Figure VI-18. Mass spectrometry was used to determine that the product of the combined biochemical and chemical reaction sequence was 93% monodeuterated. The (R)-2-deuteriosuccinate sample was found to be levorotatory.

A 5,5-dideuteriomevalonate substrate yields farnesyl pyrophosphate molecules doubly labeled with deuterium at the C-1 position via the terpenoid biosynthetic pathway (Figures VI-16 and VI-17). The biological formation of squalene from such labeled farnesyl units in the presence of NADPH then yields squalene that is trideuterated on the central methylene groups, with the single hydrogen atom on these central methylene positions being derived from NADPH.[19] Ozonolysis of squalene leads to the production of succinate from the central four carbon atoms, those carbons derived from positions one and two of the condensing farnesyl pyrophosphate molecules (Figure VI-19). The trideuteriosuccinate obtained by ozonolysis of a squalene sample biosynthesized from 5,5-dideuteriomevalonate was found to be dextrorotatory, with an optical rotatory dispersion curve that was the mirror image of that of the synthetic (R)-$(-)$-deuteriosuccinate standard. The configuration at the asymmetric center of the trideuteriosuccinate was therefore opposite to that of the reference standard, and S. As illustrated in Figure VI-19, production of (S)-2,2,3-trideuteriosuccinate by this sequence of reac-

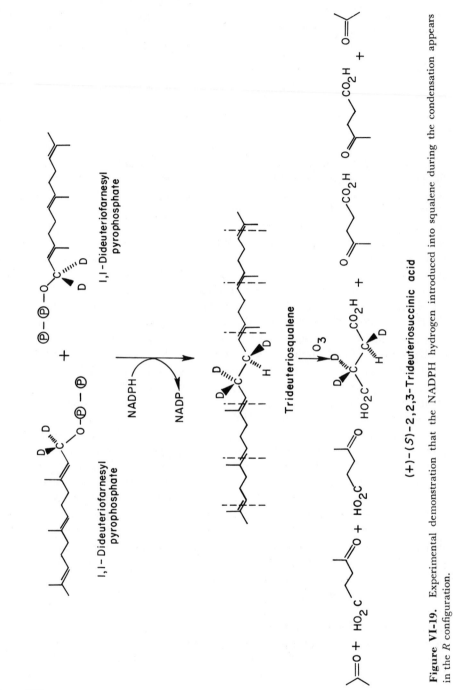

Figure VI-19. Experimental demonstration that the NADPH hydrogen introduced into squalene during the condensation appears in the R configuration.

216

tions established that the absolute configuration of the *hydrogen* introduced into squalene from NADPH in the reductive coupling reaction is R (stage 12).

It should be apparent from Figure VI-19 that an analogous procedure beginning with a 5-monodeuteriomevalonate sample of known absolute configuration could be used to answer the stereochemical question at stage 14. The necessary stereospecifically labeled 5-deuteriomevalonate was readily prepared by reducing mevaldate with NAD-^{2}H in the presence of mevaldate reductase (Mevalonate:NAD oxidoreductase) from mammalian liver. The absolute configuration of the deuterium introduced at the C-5 position of mevalonate from NAD-^{2}H in the mevaldate reductase reaction was then determined by taking advantage of the established biological stereospecificity of liver alcohol dehydrogenase (Figure V-32). Tritiated NADH (NAD-^{3}H) was used to prepare 5-tritiomevalonate, the tritiated substrate was mixed with mevalonate-4-^{14}C, and the combined material was converted enzymatically to farnesyl pyrophosphate. Alkaline phosphatase (Orthophosphoric monoester phosphohydrolase) treatment of the doubly labeled farnesyl pyrophosphate released the free terpenoid alcohol farnesol. The farnesol produced by this experimental sequence is labeled with ^{14}C at three positions (C-2, C-6, C-10) and with ^{3}H at three positions (C-1, C-5, C-9) (cf. Figures VI-16 and VI-20). The tritium labeled C-1 position of the farnesol derived by this enzymatic procedure will possess the same absolute configuration as the C-5 position of the 5-tritiomevalonate precursor. By comparing the ^{3}H/^{14}C labeling in the original farnesol (alcohol) sample with the ^{3}H/^{14}C labeling in the farnesal (aldehyde) produced by oxidation with liver alcohol dehydrogenase, it was learned that all the tritium label originally present at the C-1 farnesol position was lost during the enzyme-catalyzed oxidation.

We have described how it was established that yeast alcohol dehydrogenase stereospecifically removes the *pro-R* hydrogen of ethanol during the oxidation to acetaldehyde (cf. Chapter V, pp. 159–161). The same stereospecificity for ethanol oxidation has been found for mammalian liver alcohol dehydrogenase.[22] Furthermore, it has been determined that the liver alcohol dehydrogenase manifests the same biological stereospecificity during the oxidation of the terpenoid alcohol geraniol to geranial and transfers H_R to NAD.[23] The results obtained from the oxidation of the doubly labeled biosynthetic farnesol with the same enzyme left little doubt, therefore, that the tritium removed from the C-1 position possesses the R configuration. This established that the mevalonate produced by enzymatic reduction of mevaldate with NAD-^{3}H also possesses the R configuration at the C-5 position. This total biosynthetic and analytical sequence is outlined in Figure VI-20.

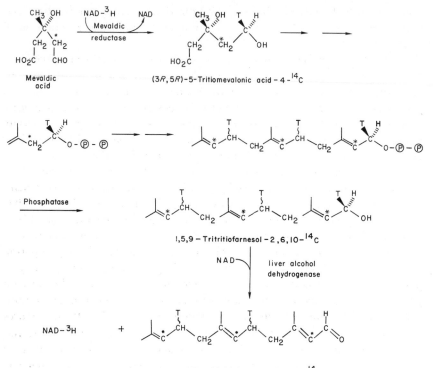

Figure VI-20. Synthesis and conversions of (3R,5R)-5-tritiomevalonate-4-¹⁴C. (Asterisk indicates the position of the ¹⁴C label).

Deuterated biosynthetic squalene was then prepared from (5R)-5-deuteriomevalonate obtained by mevaldate reduction with NAD-²H. The dideuteriosuccinate produced by ozonolysis of this biosynthetic squalene sample was found to be optically inactive. This experimental observation required that the isolated 2,3-dideuteriosuccinate be a *meso* stereoisomer. Therefore, the two central deuterated methylene carbons of the squalene sample had opposite absolute configurations. It had already been established that the *hydrogen atom* introduced into squalene *from NADPH* at stage 12 occupies the *pro-R* position at one of these methylene positions (see above). This deuterated methylene position of squalene must therefore have yielded succinate with an *S* deuterated methylene configuration. It follows that the adjacent asymmetric center of the inactive *meso*-dideuteriosuccinate, and thus of the original biosynthetic squalene molecule, must have possessed the *R* absolute configuration. The production of a deuterated methylene position in the central carbons of squalene with an R absolute configuration from a (1R)-1-deuteriofarnesyl moiety requires that the carbon bond forming process proceed with inversion at this carbon atom. At stage 14, therefore, the C-1 position of the farnesyl pyrophosphate unit which *does not* exchange hydrogen with NADPH in the coupling process *undergoes inversion* as the new carbon–carbon bond is formed (Figure VI-21).

This result also permits us to infer the stereospecificity of stage 11. Since the succinate derived from the central carbons of the biosynthetic deuterated squalene was dideuterated, it followed that neither of the deuterium labels at the C-1 positions of the (1R)-1-deuteriofarnesyl pyrophosphates undergoing condensation were lost. The hydrogen atom lost from the C-1 position of one of the condensing farnesyl units must therefore be lost from a *pro-S* paired position.*

The 4,8,12,13,17,21-hexadeuteriosqualene molecule formed biosynthetically from (5R)-5-deuteriomevalonate was also used by Popják and Cornforth to establish the stereospecificity at stages 7 and 10. The ozonolysis reaction that produces succinate from the central four carbon atoms of squalene yields four molecules of levulinate from the remainder of the molecule. The deuterated levulinates were separated from the original succinate and then oxidized with hypoiodite to yield a second sample of deuterated succinate. Mass spectrometry confirmed expectations (Figure VI-21) that this second sample of deuterated succinate was monodeuterated. The monodeuterated succinate was found to be levorotatory, like the (R)-deuteriosuccinate standard. Figure VI-21 illustrates how this result requires that the deuterated methylene positions at 4, 8, 17, and 21 of the biosynthetic hexadeuteriosqualene possess the R absolute configuration. In order to produce R methylene configurations in squalene from the (R)-1-deuterioisopentenyl pyrophosphate and the (1R)-1-deuteriogeranyl pyrophosphate intermediates [derived from the original (3R,5R)-5-deuteriomevalonate substrate], the carbon–carbon bond formation at stages 7 and 10 must occur with inversion (Figure VI-21).

The stereochemical ambiguities at stages 3, 6, and 9 of the mevalonate–squalene conversion all involve biological differentiation between enantiotopic paired methylene hydrogen atoms at the C-2 *pro*-chiral carbon center of isopentenyl pyrophosphate (Figure VI-16). This C-2 methylene of isopentenyl pyrophosphate is derived from the C-4 methylene position of mevalonate. The stereochemical course at stages 3, 6, and 9 of the biosyn-

---

* Actually, this stereochemical course at stage 11 was established by examining the ^{3}H/^{14}C label in the products obtained from (3R,5R)-5-tritiomevalonate-4-^{14}C.[24] The principle, however, is the same as that described previously. Both experiments established that the paired hydrogen at C-1 of one of the condensing farnesyl pyrophosphate units which is stereospecifically eliminated at stage 11 must be a *pro-S* hydrogen.

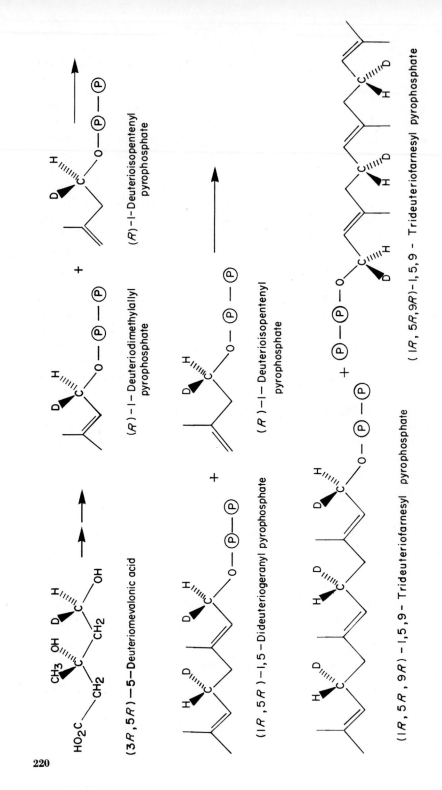

(3R, 5R) —5—Deuteriomevalonic acid

(R) —1— Deuteriodimethylallyl pyrophosphate

(R) —1— Deuterioisopentenyl pyrophosphate

(R) —1— Deuterioisopentenyl pyrophosphate

(1R, 5R) —1,5—Dideuteriogeranyl pyrophosphate

(1R, 5R, 9R) — 1,5,9 — Trideuteriofarnesyl pyrophosphate

(1R, 5R, 9R) —1,5,9 — Trideuteriofarnesyl pyrophosphate

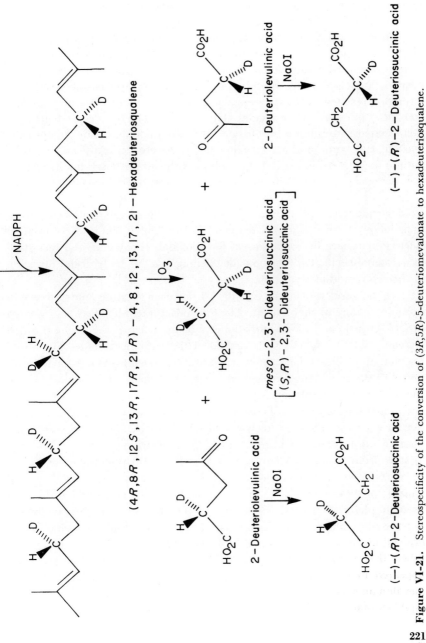

Figure VI-21. Stereospecificity of the conversion of $(3R,5R)$-5-deuteriomevalonate to hexadeuteriosqualene.

thetic pathway was therefore resolved by investigating the metabolism of 4-substituted mevalonate substrates of known absolute configurations. The preparation of the necessary substituted mevalonate was accomplished by stereospecific chemical transformations of the *trans* unsaturated hydroxy acid and the corresponding *cis* unsaturated lactone shown in Figure VI-22.

These unsaturated compounds were first converted into protected diphenylmethylamides, and then the double bonds were epoxidized by perbenzoic acid treatment. Even though two new asymmetric centers were created in each molecule by this transformation, the *syn* nature of epoxide formation resulted in the stereospecific production of only two enantiomeric epoxides from each of the amides. The two racemic epoxide products were then treated with lithium borotritide (or lithium borodeuteride) to open the epoxide ring and produce mevalonamides stereospecifically labeled with tritium (or deuterium) at C-4. Hydrolysis of the amide bonds finally yielded the necessary stereospecifically labeled 4-substituted mevalonates. The assignment of the absolute configurations in these stereospecifically synthesized mevalonates is illustrated in Figure VI-22. The assignment of configurations follows from the known geometry of the double bonds of the original starting materials, the *syn* nature of the epoxidation reaction, and the established stereospecificity of the epoxide opening reaction by hydride (inversion at the center adding hydride).

To be sure, the product yielded from each starting material was a racemic mixture of enantiomers. The *trans* unsaturated hydroxy acid yielded (3R,4R)-4-tritio and (3S,4S)-4-tritiomevalonate, whereas the *cis* unsaturated lactone yielded (3R,4S)-4-tritio and (3S,4R)-4-tritiomevalonate. The initial reaction of the biosynthetic sequence, however—the mevalonate kinase reaction (ATP:mevalonate 5-phosphotranferase)—displays a biological stereospecificity for a mevalonate substrate with the R configuration at C-3 (also designated D-mevalonate). Thus the enantiomer possessing the 3S configuration in each racemic mixture was biologically inert, and the mixture synthesized from the *trans* acid reacted as (3R,4R)-4-tritiomevalonate, with the mixture from the *cis* lactone reacting as (3R,4S)-4-tritiomevalonate.

Each of the tritium labeled mevalonate mixtures was mixed with a small amount of mevalonate-2-^{14}C as an internal control, converted to farnesyl pyrophosphate by a rat liver enzymatic preparation, and then converted to farnesol by phosphatase. Examination of the ^{14}C/^{3}H labeling ratios of the original mevalonate samples and of the biosynthetic farnesol samples established that three tritium atoms derived from (3R,4R)-4-tritiomevalonate appeared in the farnesol, but no tritium label from the (3R,4S)-4-tritiomevalonate substrate appeared in the farnesol. These experimental results establish that at stages 3, 6, and 9 the *pro-S* hydrogen at C-4 of mevalonate

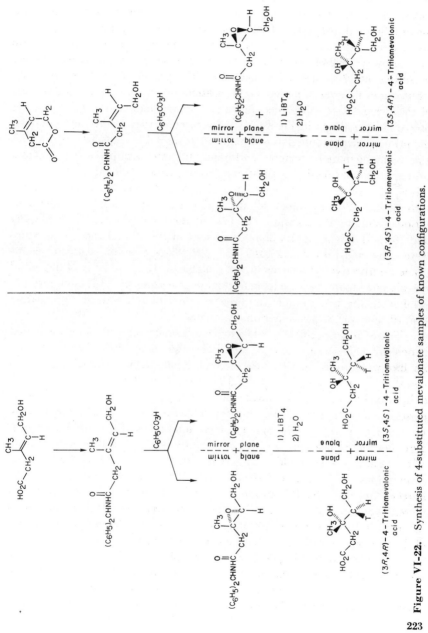

Figure VI-22. Synthesis of 4-substituted mevalonate samples of known configurations.

and the corresponding *pro-R* hydrogen at C-2 of isopentenyl pyrophosphate are stereospecifically eliminated (Figure VI-23).

Elimination of the stereochemical ambiguities at stages 1, 5, and 8 was accomplished with the aid of mevalonate substrates of known configuration substituted at the C-2 position. An examination of the 4-substituted mevalonates pictured in Figure VI-22 reveals that changing the original primary alcohol group to a carboxyl group and changing the original carboxyl group to a primary alcohol group would convert these 4-substituted mevalonate substrates into 2-substituted mevalonates. The previously synthesized 4-substituted mevalonates of known configuration are therefore potential sources of the desired C-2 substituted mevalonates. Deuterated racemic mixtures prepared as outlined in Figure VI-22 were converted to the methyl esters, the primary alcohol function was oxidized to insoluble zinc carboxylate with zinc permanganate in acetone, and finally the ester group was reduced to a primary alcohol function with lithium borohydride. As illustrated in Figure VI-24, this procedure converted the (3*S*,4*S*)-4-deuteriomevalonate isomer into (2*R*,3*R*)-2-deuteriomevalonate and the enantiomeric (3*R*,4*R*)-4-deuteriomevalonate isomer into (2*S*,3*S*)-2-deuteriomevalonate. The (3*S*,4*R*) and (3*R*,4*S*)-4-deuteriomevalonates present in the second racemic mixture were converted into the (2*S*,3*R*) and (2*R*,3*S*)-2-deuteriomevalonate enantiomers, respectively. Once again, the biological stereospecificity of the initial mevalonate kinase reaction for mevalonate possessing the (3*R*) configuration resulted in the two racemic mixtures reacting biosynthetically as separate (2*R*,3*R*) and (2*S*,3*R*)-2-deuteriomevalonates.

These stereospecifically 2-labeled mevalonate samples were then used to determine the steric course at stage 1 of the squalene biosynthetic pathway. As illustrated in Figure VI-25 for the biologically active (2*R*,3*R*)-2-deuteriomevalonate component of the (2*RS*,3*RS*) racemic mixture, an *anti* elimination of hydroxyl and carboxyl groups would yield an isopentenyl pyrophosphate possessing a vinyl deuterium atom *trans* to the methyl group (*Z* configuration, see footnote p. 148), whereas a *syn* elimination would yield isopentenyl pyrophosphate possessing a vinyl deuterium atom *cis* to the methyl group (*E* configuration). It should be apparent that the corresponding stereospecific eliminations from (2*S*,3*R*)-2-deuteriomevalonate would lead to opposite stereochemical results. Resolution of the stereochemical ambiguity at stage 1 therefore only required establishing the stereochemistries of the double bonds (*Z* or *E*) in the isopentenyl pyrophosphates produced from the metabolism of the stereospecifically 2-labeled mevalonate samples.

The conversion of isopentenyl pyrophosphate to 3,3-dimethylallyl pyrophosphate is blocked in the presence of iodoacetamide. When the two samples of 2-substituted mevalonates were separately incubated with a soluble enzyme preparation in the presence of iodoacetamide, therefore, the iso-

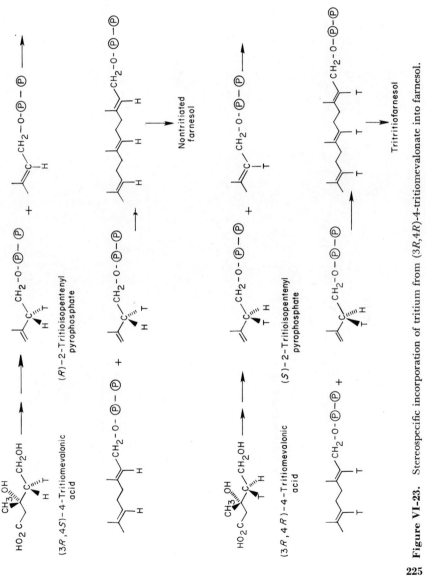

Figure VI-23. Stereospecific incorporation of tritium from $(3R,4R)$-4-tritiomevalonate into farnesol.

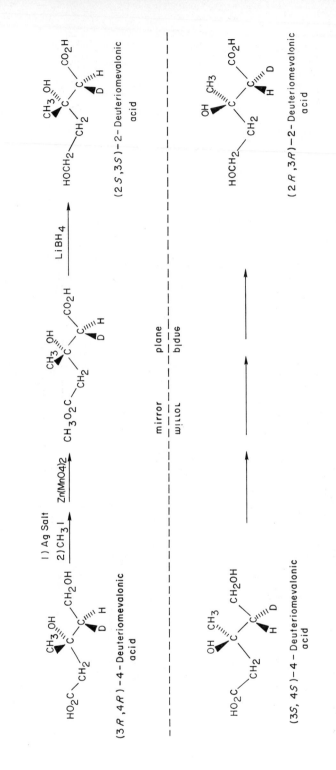

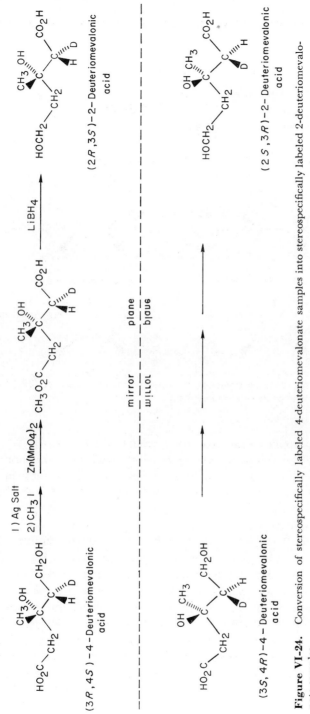

Figure VI-24. Conversion of stereospecifically labeled 4-deuteriomevalonate samples into stereospecifically labeled 2-deuteriomevalonic acid samples.

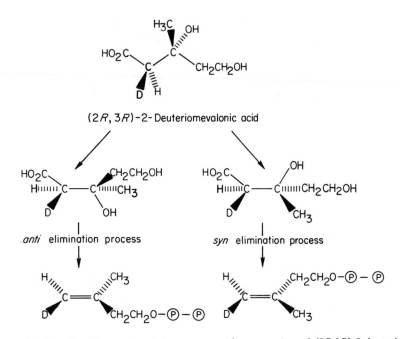

Figure VI-25. Possible results of the stereospecific conversion of $(2R,3R)$-2-deuterio-mevalonate to isopentenyl pyrophosphate.

pentenyl pyrophosphates that were produced accumulated and could be isolated. These pyrophosphate esters were hydrolyzed enzymatically to iso-pentenol with alkaline phosphatase, the isopentenol was purified, and then the relative stereochemistry around the double bonds was established by con-verting each sample of deuterated isopentenol into *cis* and *trans*-1-bromo-2-methyl-1-butenes.

The procedure used to establish the relative stereochemistry of the vinyl deuterium atom in the isopentenol samples is outlined in Figure VI-26. The vinyl deuterium atom in the isopentenyl molecules derived from $(2R,3R)$-2-deuteriomevalonate could have been either *trans* to the methyl group (*anti* elimination pathway) or *cis* to the methyl group (*syn* elimination pathway). The addition of bromine to produce 3,4-dibromo-4-deuterio-3-methyl-1-butanol is known to be a stereospecific *anti* chemical reaction. Thus the di-bromobutanol product consisted of *one or the other* of the racemic mixtures illustrated. The dibromobutanols were then treated with base in a dehydro-halogenation reaction to produce a mixture of 4-bromo-3-methyl-3-butene-1-ols. (This chemical reaction is also known to be a stereospecific *anti* process;

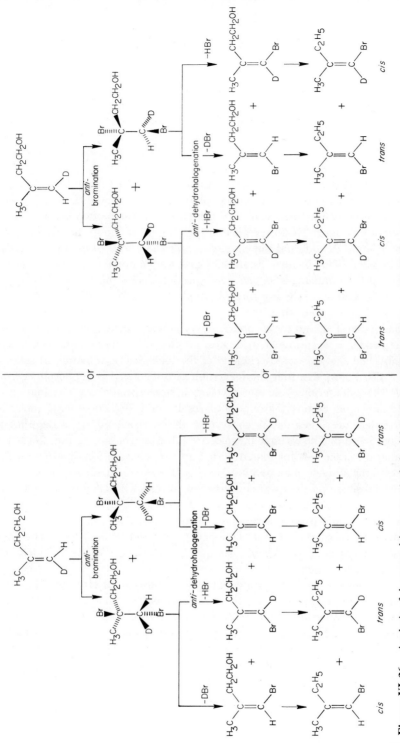

Figure VI-26. Analysis of deuterated isopentenols derived from 2-deuteriomevalonate metabolism.

229

therefore, the products are a mixture of deuterated and nondeuterated bromo-butenols, as indicated.) Finally, the alcohol group was transformed to a methyl group by reductive elimination of the tosyl ester derivatives to yield a mixture of *cis* and *trans*-1-bromo-2-methyl-1-butenes.

Standard samples of *cis* and *trans*-1-bromo-2-methyl-1-butenes had been carefully prepared by separate chemical stereospecific syntheses and thus the products derived from the biosynthetic isopentenols could be identified. It was found that the *trans*-bromomethylbutene derived from the (2*R*,3*R*)-2-deuteriomevalonate incubation was predominantly nondeuterated, whereas the *cis*-bromomethylbutene from this incubation was predominantly mono-deuterated. Figure VI-26 reveals that these results required the vinyl deuter-ium atom of the original isopentenol sample to be *trans* to the methyl group (*Z* isomer). This, in turn, establishes that the elimination at stage 1 of the biosynthetic sequence is an *anti* stereospecific process (cf. Figure VI-25).

The analysis of the isopentenol sample derived from (2S,3*R*)-2-deuterio-mevalonate showed that the final *trans*-bromomethylbutene was largely monodeuterated, and the *cis*-bromomethylbutene was largely nondeuterated. The reader is encouraged to determine how this experimental result also established the *anti* stereospecificity of the biological elimination at stage 1.

The knowledge that the formation of isopentenyl pyrophosphate from (2*R*,3*R*)-2-deuteriomevalonate involves a stereospecific *anti* elimination of hydroxyl and carboxyl groups and results in a (*Z*)-isopentenyl molecule permitted the investigators to resolve the stereochemical ambiguities at stages 5 and 8. The racemic mevalonate substrate containing the (2*R*,3*R*)-2-deuterio isomer was incubated with a soluble enzyme fraction again—this time in the absence of iodoacetamide. The biosynthetic trideuterated farnesyl pyrophosphate product was converted to farnesol with alkaline phosphatase, and the farnesol was ozonized to yield deuterated levulinate. The levulinate was then oxidized with hypoiodite to succinate. Analysis of this succinate sample revealed that it was monodeuterated and levorotatory. The com-bined biosynthetic and chemical sequence described therefore yielded (*R*)-2-deuteriosuccinate.

Isopentenyl pyrophosphate molecule contains an unsaturated $X_{aa}=Y_{bc}$ *pro*-chiral grouping in which the two faces of the double bond are geo-metrically distinct (cf. Figure V-22 and related discussion). We can therefore expect that the carbon bond forming reactions at stages 5 and 8 will be biolog-ical stereospecific reactions in which the new bonds form by addition of the allylic groups to either the *re,re* or the *si,si* face of the isopentenyl pyrophos-phate substrate. As illustrated in Figure VI-27, in the case of the stereo-specifically deuterated isopentenyl pyrophosphate substrate, these two stereo-specific routes can be differentiated because they yield contrasting stereoisomeric products. Figure VI-27 shows how the production of (R)-2-deuterio-succinate from the deuterated biosynthetic farnesol product requires that the carbon bond forming processes at stages 5 and 8 involve addition to the 3si,4si face of the stereospecifically deuterated (Z)-isopentenyl pyrophosphate intermediate. This corresponds to a stereospecific addition to the si,si face on the nondeuterated substrate and thus establishes the stereospecificity at stages 5 and 8 of squalene biosynthesis.

Of the 14 stages of stereochemical ambiguity in the mevalonate to squalene conversion originally designated by Popják and Cornforth, only the stereospecificity of stage 2 remains undefined. The reaction catalyzed by iso-pentenyl pyrophosphate isomerase (Isopentenylpyrophosphate Δ^3-Δ^2-iso-merase) involves the addition of hydrogen to an unsaturated methylene group to form a new methyl group. This process can be schematically represented as

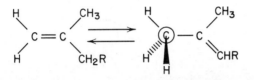

This analysis emphasizes that the isopentenyl pyrophosphate isomerase reaction is a representative of a second class of methyl–methylene intercon-version reactions. It is related to the methyl–methylene interconversion re-actions discussed earlier in this chapter (p. 193). The essential difference is that the methylene function involved in this second class of methyl–methylene interconversions is unsaturated, and thus the stereospecificity can be ana-lyzed in terms of a re or si addition to the geometrically nonequivalent faces of an unsaturated pro-chiral grouping.

The manner of establishing the biological stereospecificity at stage 2 should be apparent from the relationship that exists between these two classes of methyl–methylene interconversions. In an isopentenyl pyrophosphate molecule stereospecifically labeled at the methylene function with hydrogen and deuterium, the addition of tritium from the re or from the si face will yield methyl groups of opposite chiralities (Figure VI-28). As described earlier, the appropriate labeled isopentenyl pyrophosphate substrates of known stereochemistry have been prepared. Since methods permitting the determination of the absolute configuration of chiral methyl groups have also been developed (e.g., via the malate synthetase–fumarase assay of chiral acetates described earlier), it is highly probable that the stereochemical course at stage 2 of the mevalonate to squalene conversion will be established in the near future.

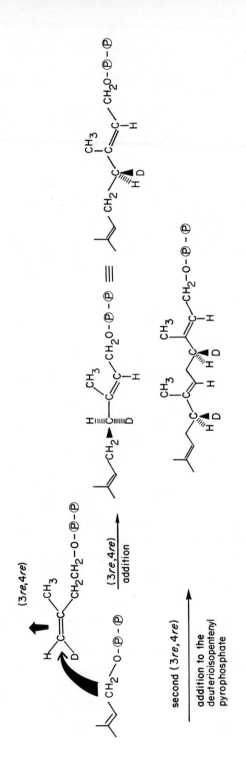

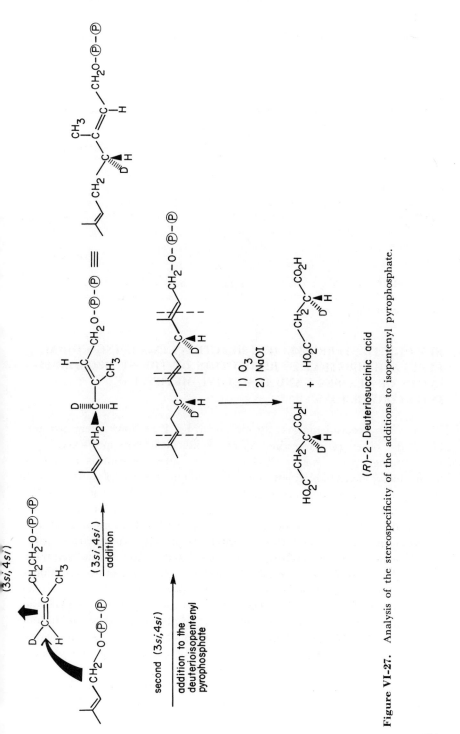

Figure VI-27. Analysis of the stereospecificity of the additions to isopentenyl pyrophosphate.

233

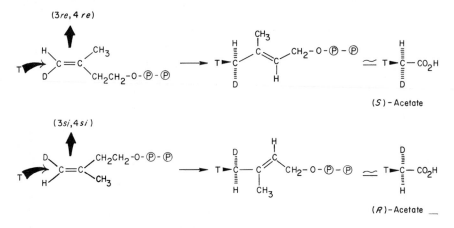

Figure VI-28. Potential method of resolving remaining stereochemical ambiguity in squalene biosynthesis pathway (stage 2). Note added in proof: Cf. Ref. 71 wherein a *re,re* pathway is established.

BIOLOGICAL STEREOSPECIFIC REACTIONS INVOLVING CHIRAL PYRUVATE SUBSTRATES: RESOLUTION OF THE STEREOSPECIFICITIES OF CLASSES I AND II METHYL–METHYLENE INTERCONVERSIONS

Recent experimental results obtained by Rose have increased our knowledge of several interesting aspects of biological stereospecificity.[25]

Although the class II methyl–methylene interconversion of the mevalonate to squalene biosynthetic pathway remains a stage of undetermined stereospecificity, experiments recently reported by Rose have established the stereospecificity of an analogous class II biological methyl–methylene interconversion. Rose's investigations have also resulted in the formation of new biochemical substrates with chiral methyl groups of known stereochemistry, the (R) and (S)-3-deuterio-3-tritiopyruvates. Finally, Rose has utilized these chiral pyruvate substrates to establish the stereospecificities of two additional biological methyl–methylene interconversions of class I.

As illustrated in Figure VI-29, the pyruvate kinase (ATP:pyruvate phosphotransferase) reaction converts an unsaturated methylene into a methyl function. Therefore, like the isopentenyl pyrophosphate isomerase reaction, the pyruvate kinase reaction may be considered as a biological methyl–methylene interconversion of class II.

In Chapter V it was pointed out that phosphoenolpyruvate is an important biological intermediate, possessing geometrically nonequivalent *re,re* and *si,si* faces (cf. Figure V-22 and related discussion). The stereospecificity of

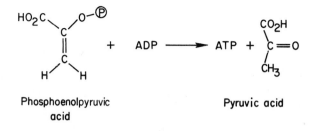

| | | CO₂H |
| Phosphoenolpyruvic acid | | Pyruvic acid |

Figure VI-29.   Methyl–methylene interconversion catalyzed by pyruvate kinase.

the methylene–methyl conversion catalyzed by pyruvate kinase can thus be analyzed in terms of a *re* or *si* addition of a hydrogen atom to the unsaturated *pro*-chiral grouping of phosphoenolpyruvate. By taking advantage of the previously established stereospecificities of the glucose phosphate isomerase (D-Glucose-6-phosphate ketol-isomerase) and the enolase (2-Phospho-D-glycerate hydrolyase) reactions (cf. pp. 250–254 following), Rose was able to prepare the two specifically labeled phosphoenolpyruvate substrates presented in Figure VI-30.[25] As shown for the substrate possessing the *cis* arrangement of tritium and phosphate (*Z* isomer), the addition of a hydrogen to the *re,re* or to the *si,si* face of the methylene carbon atom in the pyruvate kinase reaction will lead to pyruvate molecules with chiral methyl groups of opposite absolute configurations.* The pyruvate products were oxidized to acetate with H_2O_2, and the acetate was purified by ion-exchange chromatography. The chiralities of the acetate products were then determined by the malate synthetase–fumarase analytical sequence. Beginning with the (*Z*)-phosphoenolpyruvate substrate, a malate sample was ultimately obtained in which only 16% of the tritium label was retained in fumarate-malate after incubation with fumarase. This malate was therefore derived by a malate synthetase-catalyzed condensation of (*S*)-2-deuterio-2-tritio-acetate (cf. pp. 205–208). Beginning with the (*E*)-phosphoenolpyruvate substrate, a malate sample was obtained in which 68% of the tritium label was retained in fumarate-malate after incubation with fumarase. The latter malate intermediate was therefore derived by the condensation of (*R*)-2-deuterio-2-tritioacetate.

As illustrated in Figure VI-30, the stereospecific production of (*S*)-2-deuterio-2-tritioacetate from the phosphoenolpyruvate sample with a *cis* arrangement of tritium and phosphate requires a stereospecific addition of hydrogen to the 2*si*,3*re* face of the unsaturated *pro*-chiral methylene carbon

* Again of course, only those substrate molecules which actually contain tritium atoms at the methylene carbons will be converted into chiral pyruvate molecules.

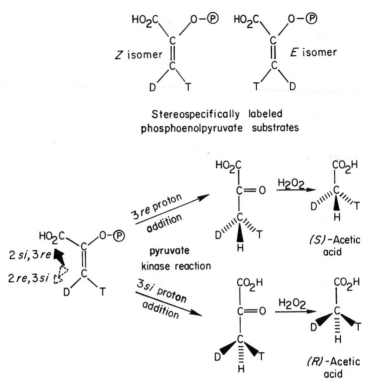

Figure VI-30.   Possible stereospecific results of the pyruvate kinase-catalyzed process.

in the pyruvate kinase-catalyzed transformation. The reader should convince himself that the stereospecific production of (R)-2-deuterio-2-tritioacetate from the substrate with a *trans* arrangement of tritium and phosphate also requires a stereospecific 2si,3re addition of hydrogen in the pyruvate kinase reaction. These stereospecific additions correspond to the addition of hydrogen to the si,si enantiotopic face of an unlabeled phosphoenolpyruvate substrate.

The foregoing experiments not only established the stereospecificity of a biological methyl–methylene interconversion of class II, they also resulted in the production of pyruvate samples of known chirality. The (R) and (S)-3-deuterio-3-tritiopyruvates formed and characterized as outlined in Figure VI-30 possess chiral methyl groups. These enantiomeric pyruvates are therefore additional examples of biologically active substrates possessing chiral methyl groups of established absolute configurations and should be considered with the chiral acetates described earlier in this chapter.

Rose used these chiral pyruvate substrates to establish the stereospecificities of the two additional biological methyl–methylene interconversions (class I) pictured in Figure VI-31.[25]

To establish the stereospecificity of the pyruvate carboxylase reaction [Pyruvate:carbon-dioxide ligase (ADP)], each of the chiral pyruvate substrates was carboxylated to oxaloacetate with pyruvate carboxylase in the presence of malate dehydrogenase. Since it was known that the pyruvate carboxylase reaction manifests a normal kinetic isotope effect ($k_H/k_T = 4.2$), the absolute stereochemistry of the tritium labeling of the malate product could be used to establish the stereospecificity of the carboxylation process (cf. pp. 199–201). Once again, the fumarase reaction was utilized to establish the stereochemistry of the tritium label in the malate products; the tritium lost or exchanged from the C-3 position of malate during the fumarase-catalyzed reaction will possess the R absolute configuration (cf. p. 205). The malate sample isolated after the coupled enzymatic conversions of the (S)-3-deuterio-3-tritiopyruvate lost 78% of the tritium label upon incubation with fumarase. This malate sample was therefore predominantly ($2S,3R$)-3-deuterio-3-tritiomalate. In contrast, the malate sample isolated after the coupled conversions of the (R)-3-deuterio-3-tritiopyruvate lost only 36% of the tritium label upon incubation with fumarase. Thus the latter malate sample was predominantly ($2S,3S$)-3-deuterio-3-tritiomalate.

Figure VI-32 illustrates how, *given the existence of a normal kinetic isotope effect* ($k_H/k_D > 1$) during the pyruvate carboxylase reaction, the formation of a malate sample with predominant ($3R$) tritium labeling from (S)-3-deuterio-3-tritiopyruvate requires that the carboxylation reaction occur with

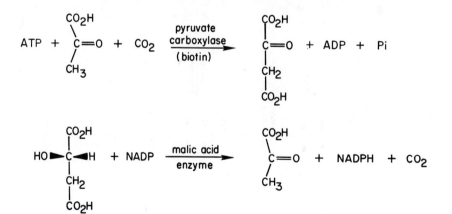

Figure VI-31. Biological methyl–methylene interconversions (class I) involving pyruvate.

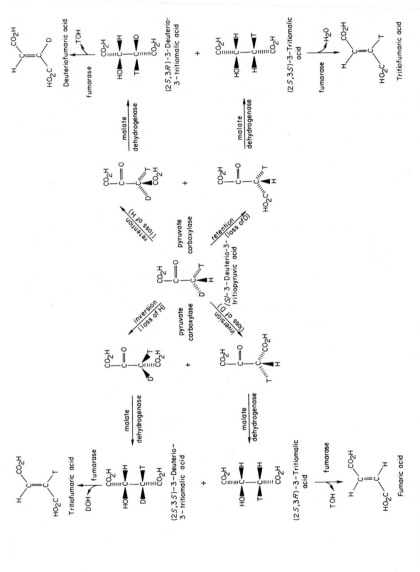

Figure VI-32. Possible stereospecific conversions of (S)-3-deuterio-3-tritiopyruvate by pyruvate carboxylase and malate dehydrogenase.

retention of configuration at the C-3 (methyl) carbon of pyruvate. Similarly, we can show that the formation of a malate sample with predominant (3S) tritium labeling from (R)-3-deuterio-3-tritiopyruvate also requires that the methyl–methylene interconversion catalyzed by pyruvate carboxylase occur with *retention* of configuration.

As the next step in his investigations of biological stereospecific carboxylation reactions, Rose utilized the (2S,3S) and (2S,3R)-deuterio-3-tritiomalate samples prepared and identified during the course of the experiments just described to determine the biological stereospecificity of the malic acid enzyme [L-Malate: NADP oxidoreductase (decarboxylating)] from *Escherichia coli* and pigeon liver.[25] As illustrated in Figure VI-33, both the (3R) and (3S)-3-tritiomalate samples were converted to pyruvate and carbon dioxide by incubation with the malic acid enzyme and NADP. By carrying out these incubations in H_2O, the methylene–methyl interconversions catalyzed by the malic acid enzymes converted the chiral 3-deuterio-3-tritiomethylene groups of the malate substrates into chiral 3-deuterio-3-tritiomethyl groups in the pyruvate products.

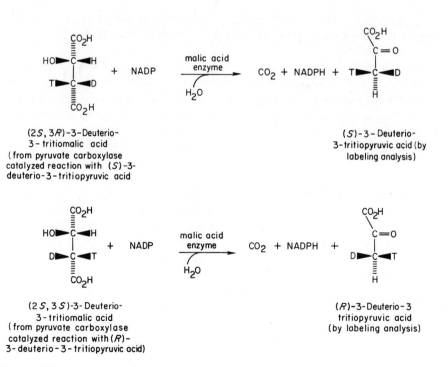

(2S, 3R)-3-Deuterio-
3-tritiomalic acid
(from pyruvate carboxylase
catalyzed reaction with (S)-3-
deuterio-3-tritiopyruvic acid

(S)-3-Deuterio-
3-tritiopyruvic acid (by
labeling analysis)

(2S, 3S)-3-Deuterio-
3-tritiomalic acid
(from pyruvate carboxylase
catalyzed reaction with (R)-
3-deuterio-3-tritiopyruvic acid)

(R)-3-Deuterio-3
tritiopyruvic acid
(by labeling analysis)

Figure VI-33. Stereospecificity of the malic acid enzyme-catalyzed conversion of malate to pyruvate.

The chiral pyruvate products from each of these incubations were isolated and converted into fumarates via the sequence of reactions catalyzed by pyruvate carboxylase, malate dehydrogenase, and fumarase. The chiralities of these pyruvate samples were then established by comparing the tritium labeling results from this sequence of reactions with the tritium labeling results from the same sequence carried out with the previously identified (R) and (S)-3-deuterio-3-tritiopyruvate samples (cf. Figure VI-32). The $(2S,3R)$-3-deuterio-3-tritiomalate sample yielded a pyruvate sample whose behavior in the pyruvate carboxylase–malate dehydrogenase–fumarase sequence indicated that it was predominantly (S)-3-deuterio-3-tritiopyruvate. The $(2S,3S)$-3-deuterio-3-tritiomalate sample yielded a pyruvate sample whose behavior in the analytical sequence suggested it was predominantly (R)-3-deuterio-3-tritiopyruvate. As indicated in Figure VI-33, these results establish that the methylene–methyl interconversion catalyzed by the malic acid enzyme occurs stereospecifically with *retention* of configuration.

BIOLOGICAL STEREOSPECIFICITIES TOWARD THE PAIRED METHYLENE HYDROGENS AT C-4' OF NADH AND NADPH

That biological stereospecificity would result in the enzymatic differentiation of both enantiotopic and diastereotopic paired hydrogens was first established during the course of investigations with yeast alcohol dehydrogenase preparations.[26] The fundamental significance of these investigations of alcohol dehydrogenase, and the fact that these studies were undertaken early in the 1950s qualifies them to be ranked with the investigations of the citrate synthetase–aconitase reactions as "classical" demonstrations of biological stereospecificity toward chemically like, paired groups. The experiments that proved yeast alcohol dehydrogenase differentiates between enantiotopic paired hydrogens at the *pro*-chiral C-1 position of ethanol are described in Chapter V. The same series of investigations also established that the yeast alcohol dehydrogenase system differentiates between the diastereotopic paired hydrogens at the C-4' position of the dihydronicotinamide ring of NADH.

Biochemically deuterated NADH molecules were prepared by reducing NAD with 1,1-dideuterioethanol in the presence of a yeast alcohol dehydrogenase preparation. (It had previously been learned that direct transfer of deuterium from the C-1 position of ethanol to the nicotinamide moiety occurred and that protons from the solvent were not incorporated into the reduced pyridine nucleotide.[27]) This biosynthetic sequence therefore produced a reduced pyridine nucleotide bearing one deuterium atom at C-4' of the dihydronicotinamide ring (NAD-^{2}H). When this biosynthetic NAD-^{2}H was used to reduce acetaldehyde in the presence of alcohol dehydrogenase, it was found that the entire deuterium label was stereospecifically transferred to yield a monodeuterated ethanol product and nondeuterated NAD. Chemically monodeuterated NAD-^{2}H was prepared by hydrosulfite reduction of NAD in D_2O (99.98% deuterated). When this chemosynthetic NAD-^{2}H was used to reduce acetaldehyde in the presence of alcohol dehydrogenase, the NAD product retained 0.44 atom of deuterium per molecule. * Thus we know that the yeast alcohol dehydrogenase enzyme stereospecifically transferred deuterium from C-1 of ethanol to yield NAD-^{2}H labeled with deuterium in only one unique diastereotopic position. The enzymatic reoxidation of the biosynthetic NAD-^{2}H catalyzed by alcohol dehydrogenase then stereospecifically transferred the same diastereotopic deuterium label to acetaldehyde. In contrast, the chemically prepared NAD-^{2}H substrate was labeled with deuterium in both diastereotopic positions in nearly equal amounts, and only deuterium label at C-4' of the dihydronicotinamide ring of the appropriate absolute configuration was removed during the oxidation catalyzed by alcohol dehydrogenase.

Originally the two enzymatically differentiable methylene hydrogens of the reduced nicotinamide ring of NADH were designated as α and β. However, since α and β were also used to designate the stereochemistry at the nucleosidic (ribose-nicotinamide) bond of NAD and NADH, these diastereotopic paired methylene hydrogens have come to be designated as A and B. The hydrogen removed from NADH (or added to NAD) in the yeast alcohol dehydrogenase reaction described previously is designated the "A" hydrogen. The diastereotopic paired hydrogen is designated as "B." Yeast alcohol dehydrogenase therefore can be said to belong to the stereospecific "A class" of dehydrogenase enzymes.

Similar experiments with other enzymatic preparations established that biological stereospecificity toward the paired hydrogens at C-4' of the reduced nicotinamide ring of NADH and NADPH is not a unique property of yeast alcohol dehydrogenase. It has been found that this type of biological stereospecificity is a general enzymatic phenomenon. Since it was possible to stereospecifically prepare A-NAD-^{2}H (e.g., by reduction of NAD with

---

* This result indicates that the foregoing chemical reduction manifests a slight degree of stereoselectivity and yields a mixture of monodeuterated NAD-^{2}H molecules containing 56% of the diastereomer with deuterium in the enzymatically active (with alcohol dehydrogenase) configuration. Chemical oxidations of NAD-^{2}H samples have also been found to display stereoselectivity and tend to remove deuterium from one diastereotopic position preferentially.[28] Well before 220-MHz nmr, therefore, these observations established the chemical nonequivalence of the paired hydrogens at C-4' of the dihydronicotinamide ring of NADH (cf. pp. 46–50).

1,1-dideuterioethanol in the presence of alcohol dehydrogenase) and B-NAD-^{2}H (e.g., by reduction of deuterated NAD with unlabeled ethanol in the presence of alcohol dehydrogenase), the relative biological stereospecificity observed in different enzymatic reactions could also be established to be of class A (same stereospecificity as yeast alcohol dehydrogenase) or of class B (opposite stereospecificity to yeast alcohol dehydrogenase). In *Molecular Asymmetry in Biology*, Volume II,[29] there is a listing of more than 40 separate enzymatic oxidoreductase systems that have been found to manifest biological stereospecificity toward the paired diastereotopic hydrogens of the dihydronicotinamide ring of the pyridine nucleotides. In many instances the stereospecificity of these enzymatic reactions has been investigated with preparations from different organisms and with more than one substrate. As the reader may have been anticipated from our discussion of the two citrate synthetases of opposite biological stereospecificities, not all the oxidoreductase reactions studied have the same relative specificity as the initially classified yeast alcohol dehydrogenase. Many of the oxidoreductase enzymes react stereospecifically with the paired B hydrogen at C-4′ of NADH or NADPH and are therefore classified as B stereospecific systems. The established A or B stereospecificities of the enzymatic systems investigated are listed in tables compiled by Bentley.[29]

The summarized results of biological stereospecificity toward the paired hydrogens at C-4′ of NADH and NADPH fail to reveal any particular reaction preference for the A or B hydrogen. Thus some enzymatic reactions involving the oxidations of hydroxyl groups to carbonyl groups proceed with complete A stereospecificity and others with complete B stereospecificity. Similarly, enzymatic oxidations of carbonyl groups proceed with A or with B stereospecificity depending on the specific enzyme involved. Various enzymes involved in respiratory-linked dehydrogenase processes also belong to both stereospecific classes. Certain generalizations, however, can be made regarding this area of biological stereospecificity.

1. A given enzyme will be stereospecific for either the A or B paired hydrogen and will, therefore, belong to a single classification.

2. In the case of those particular reactions in which either NADH or NADPH may function as the coenzyme, the stereospecificity is the same with both coenzymes.

3. If a given enzyme can catalyze the oxidation-reduction of a variety of substrates, the stereospecificity of the hydrogen transfer will be the same for all the substrates (cf., the discussion on p. 217).

4. With the exception of some of the dehydrogenase enzymes involved in electron transport processes, the stereospecificity of a particular reaction is fixed and does not vary with the source of the enzyme preparation.

Once again, this biological stereospecificity toward chemically like, paired groups was observed and the relative stereochemistry was established several years before the absolute biological stereospecificities could be determined. The absolute configurations of the A and B paired hydrogens at C-4' of the dihydronicotinamide ring of NADH were not definitively determined until 1964,[30] even though the original observations of biological stereospecificity toward these paired hydrogens had been reported in 1953.[26] As part of the investigation of the stereospecificity of the mevalonate to squalene conversions in 1964, Cornforth et al. prepared NAD-^{2}H samples labeled with deuterium in the A and in the B positions. Each sample then was chemically degraded to yield succinates stereospecifically labeled with deuterium in the methylene group. The chemical degradation involved treating the reduced pyridine nucleotide with acetic acid–methanol and then oxidizing the product sequentially with ozone and perbenzoic acid. Although the intermediates in this transformation were not completely characterized, the process did produce succinate in about 18% overall yield. The degradation sequence insured that methylene groups of the succinate yielded by this process were derived from the vital C-4' and C-5' positions of the NADH molecule. When A-NAD-^{2}H and B-NAD-^{2}H samples were degraded by the foregoing procedure, it was found that the succinates produced were largely monodeuterated and optically active. The monodeuterio-succinate sample derived from A-NAD-^{2}H was levorotatory; that from B-NAD-^{2}H was dextrorotatory.

Earlier in this chapter, we described how Cornforth et al. prepared an (R)-2-deuteriosuccinate standard (cf., Figure VI-18). Since the standard (R)-2- deuteriosuccinate was levorotatory, direct comparison showed that the monodeuteriosuccinate derived from A-NAD-^{2}H was (R)-2-deuteriosucci-nate. In turn, as illustrated in Figure VI-34, this established that the biosynthetic A-NAD-^{2}H corresponds to the absolute configuration of $(4'R)$-NAD-^{2}H, and B-NAD-^{2}H corresponds to the absolute configuration of $(4'S)$-NAD-^{2}H. The paired hydrogen originally designated A can now be assigned the H_R absolute configuration, whereas the B methylene hydrogen can be assigned the H_S configuration. The enzymes listed in the A stereo-specific classification (e.g., in Tables I and II of Ref. 29) should now be listed as enzymes stereospecific for H_R of NADH or NADPH and those in the B stereospecific classification as H_S stereospecific enzymes. Unfortunately, the widespread usage of the historical A and B relative stereoclassifications has hampered adoption of the more meaningful *pro*-chiral designations. We urge the reader to describe the biological stereospecificity of enzymes toward the paired methylene hydrogens at C-4' of the reduced nicotinamide ring of NADH and NADPH in terms of the more informative H_R and H_S absolute configurational designations.

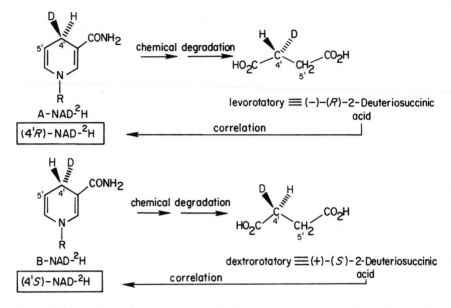

Figure VI-34. Determination of the absolute configurations of A-NAD-²H and B-NAD-²H.

BIOLOGICAL STEREOSPECIFICITIES IN SACCHARIDE PHOSPHATE KETOL–ISOMERASE REACTIONS: PREPARATION OF STEREO-SPECIFICALLY LABELED PHOSPHOENOLPYRUVATE SUBSTRATES VIA KETOL–ISOMERASE AND ENOLASE REACTIONS; MECHANISTIC INFORMATION REGARDING THE KETOL–ISOMERASE REACTIONS

The biological stereospecificities manifest in various saccharide phosphate ketol-isomerase reactions have been intensively investigated. This aspect of biological stereospecificity can be readily understood and anticipated in terms of the principles developed in the earlier chapters of this book. Examples of biological stereospecificities in ketol-isomerase reactions, however, may also be used to demonstrate how biochemical processes of established stereospecificity can yield valuable labeled substrates required for further investigations. In addition, the studies of ketol-isomerase reactions demonstrate how investigations of biological stereospecificity can yield results of mechanistic importance.

The D-ribulose 5-phosphate molecule in Figure VI-35 reveals the vital structural features of the ketose phosphates which are involved in ketose-aldose isomerizations (saccharide phosphate ketol-isomerase reactions). The C-1 position (hydroxymethyl carbon) of D-ribulose 5-phosphate is a *pro*-chiral

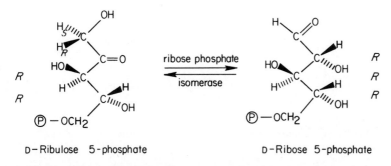

D−Ribulose   5−phosphate   D−Ribose   5−phosphate

Figure VI-35.   Ribose phosphate isomerase reaction.

carbon and bears chemically like, paired hydrogen atoms. In contrast to the enantiotopic relationship of the paired hydrogens of the C-1 position (hydroxymethyl carbon) of ethanol, however, the chiral centers at C-3 and C-4 of D-ribulose result in a diastereotopic relationship between the paired hydrogen atoms at C-1 of D-ribulose 5-phosphate.

As indicated in Figure VI-35, ribose phosphate isomerase (D-Ribose-5-phosphate ketol-isomerase) catalyzes the interconversion of the ketose D-ribulose 5-phosphate and the aldose D-ribose 5-phosphate. During the iso-merization of D-ribulose 5-phosphate to D-ribose 5-phosphate, an additional chiral center is generated at C-2. That aspect of biological stereospecificity which leads to the production of a single stereoisomer uniquely produces a D (or R) absolute configuration at this new chiral center to yield the D-ribose 5-phosphate molecule shown. The alternative D-arabinose 5-phosphate structure, therefore, is not formed in this particular isomerase reaction. Obviously, the isomerization of D-ribulose 5-phosphate to D-ribose 5-phos-phate requires the loss (or transfer, see below) of a hydrogen atom originally present at the C-1 position of D-ribulose 5-phosphate. Additionally therefore, biological stereospecificity will lead to the removal from C-1 of only one of the chemically like, paired hydrogens of the D-ribulose 5-phosphate.

The existence of the anticipated biological stereospecificity toward the paired hydrogen atoms at C-1 of D-ribulose 5-phosphate was established by incubating D-ribose 5-phosphate in tritiated water in the presence of ribose phosphate isomerase. When equilibrium had been established, the enzy-matically formed D-ribulose 5-phosphate was isolated and was found to have incorporated one atom of tritium per ribulose molecule.[31,32] The chemical and biochemical reactions indicated in Figure VI-36 were then utilized to establish that the D-1-tritioribulose 5-phosphate formed possessed the R absolute configuration at the tritiated C-1 position.[32]

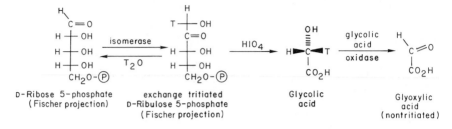

Figure VI-36. Determination of absolute stereospecificity of the ribose phosphate isomerase reaction.

Periodate oxidation of the biochemically labeled D-ribulose 5-phosphate yielded tritiated glycolate derived from the C-1 and C-2 ribulose carbon atoms. After purification, this tritiated glycolate was enzymatically oxidized to glyoxylate with glycolic acid oxidase (Glycolate:NAD oxidoreductose). The final glyoxylate produced by this sequence lacked tritium labeling. The glycolic acid oxidase reaction is analogous to the previously described alcohol dehydrogenase reaction and manifests biological stereospecificity for one of the enantiotopic methylene hydrogens. Since it was known that glycolic acid oxidase stereospecifically removes the *pro-R* paired hydrogen atom from glycolate in the oxidation to glyoxylate,* the above-mentioned experimental results established that the tritiated glycolate, and hence the D-1-tritio-ribulose 5-phosphate, possessed the *R* absolute configuration at the tritiated

---

* The biological stereospecificity of glycolic acid oxidase for the H_R methylene hydrogen was originally inferred from the observation that glycolic oxidase would oxidize L-lactate but not D-lactate to pyruvate.[33] L-Lactate possesses a hydrogen that corresponds geometrically to the H_R paired hydrogen of glycolate, whereas the inert D-lactate molecule possesses a methyl group in this corresponding position (Figure VI-37).

This stereochemical conclusion was supported by the observation that the tritiated glycolate prepared by treating glyoxylate with muscle L-lactate dehydrogenase (L-lactate: NAD oxidoreductase) in the presence of NAD-³H yielded unlabeled glyoxylate upon subsequent oxidation with glycolic acid oxidase.[34] Again, since the reduction of pyruvate to L-lactate by lactate dehydrogenase must introduce a hydrogen atom into the position that corresponds geometrically to the H_R position in glycolate, it was assumed that the tritium introduced into glycolate by L-lactate dehydrogenase and subsequently removed by glycolic acid oxidase possessed the *R* configuration (Figure VI-37). The biological stereospecificity of glycolic acid oxidase was definitively established by X-ray and neutron-diffraction investigations of the 2-deuterioglycolate produced by the L-lactate dehydrogenase-catalyzed reduction of 2-deuterioglyoxylate. The 2-deuterioglycolate formed in this enzymatic process was found to be (*S*)-2-deuterioglycolate.[35] Thus, as illustrated in Figure VI-37, the tritium introduced into glycolate by lactate dehydrogenase reduction with NAD-³H was indeed *R*. Since this tritium was stereospecifically removed in the glycolic acid oxidase reaction, the H_R biological stereospecificity of this enzyme was established.

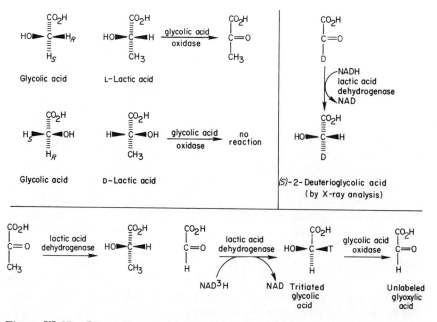

Figure VI-37. Determination of the biological stereospecificity of glycolic acid oxidase.

position (Figure VI-36). These results showed that ribose phosphate isomerase does differentiate between the diastereotopic paired hydrogens at C-1 of the ketose D-ribulose 5-phosphate and that, during the catalyzed conversion of D-ribose 5-phosphate to D-ribulose 5-phosphate, the hydrogen introduced onto the C-1 carbon atom is the *pro-R* hydrogen. Under equilibrating conditions this *pro-R* hydrogen is derived from protons in the aqueous solvent, and during the reverse conversion of D-ribulose 5-phosphate to D-ribose 5-phosphate, this *pro-R* hydrogen is stereospecifically lost into the solvent.

Since the structural features of D-ribose 5-phosphate are common to other ketose phosphates involved in intermediary metabolism, manifestations of biological stereospecificity analogous to those described for the ribose phosphate isomerase reaction will be observed in other ketol-isomerase systems. However, although the ketol-isomerase enzymes always differentiate between the paired methylene hydrogen atoms of the C-1 hydroxymethyl grouping, the absolute stereospecificity of this biological differentiation varies with the enzyme. Certain ketol-isomerases specifically remove the *pro-R* paired hydrogen atom; others may remove the diastereotopic *pro-S* hydrogen. The situation is analogous to the previously discussed (*re*) and (*si*)-citrate synthetases and the H$_R$ (A) and H$_S$ (B) dehydrogenases.

This feature of the biological stereospecificity of ketol-isomerases was clearly demonstrated in experiments carried out by Topper with glucose phosphate isomerase and mannose phosphate isomerase (D-Mannose-6-phosphate ketol-isomerase).[36] Glucose phosphate isomerase catalyzes the interconversion of the ketose fructose 6-phosphate and the aldose glucose 6-phosphate. Mannose phosphate isomerase catalyzes the interconversion of the same fructose 6-phosphate and the aldose mannose 6-phosphate, as shown in Figure VI-38. These separate ketol-isomerase enzymes thus manifest opposite biological stereospecificities with regard to the new *chiral center* generated in the ketose to aldose conversion. Glucose phosphate isomerase stereospecifically converts the ketone group of fructose into a secondary alcohol group of the *R* absolute configuration. Mannose phosphate isomerase stereospecifically converts the ketone group of fructose into a secondary alcohol group of the *S* absolute configuration. Considering the reverse aldose–ketose conversions, these two enzymes can be characterized as possessing contrasting steric specificities toward the aldose C-2 chiral center. Glucose phosphate isomerase stereospecifically converts glucose 6-phosphate into fructose 6-phosphate, whereas the mannose phosphate isomerase enzyme possesses a steric specificity for the diastereomeric aldose and converts mannose 6-phosphate into fructose 6-phosphate.

When Topper incubated 1-deuterioglucose 6-phosphate with glucose phosphate isomerase in H_2O under equilibrating conditions, the deuterium label was not lost from the hexoses to the solvent. This observation demonstrated that the hydrogen eliminated in the fructose phosphate–glucose phosphate conversion is the same hydrogen stereospecifically introduced into the C-1 position of fructose 6-phosphate during the glucose phosphate–fructose phosphate isomerization. Owing to this biological stereospecificity,

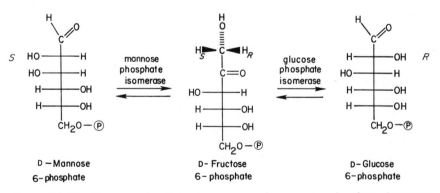

Figure VI-38. Glucose phosphate isomerase and mannose phosphate isomerase reactions.

therefore, even multiple glucose–fructose interconversions catalyzed by glucose phosphate isomerase will not lead to the loss of the hydrogen (deuterium) originally present on the C-1 aldehydic carbon of glucose. The addition of mannose phosphate isomerase to a glucose phosphate isomerase incubation, however, did lead to the loss of the deuterium label originally present as 1-deuterioglucose 6-phosphate into the solvent. Taken together, these experimental observations demonstrate that the ketol-isomerases differentiate between the paired hydrogens at the C-1 position of fructose 6-phosphate and that glucose phosphate isomerase and mannose phosphate isomerase manifest opposite absolute biological stereospecificity toward these paired hydrogens. The hydrogen (deuterium) originally present at C-1 of glucose 6-phosphate is stereospecifically retained in the glucose phosphate isomerase-catalyzed interconversions of fructose 6-phosphate; but this hydrogen (deuterium) is stereospecifically lost to the solvent in the mannose phosphate isomerase-catalyzed interconversions of fructose 6-phosphate.

Rose and O'Connell established that the glucose phosphate isomerase possesses an absolute biological stereospecificity for the *pro-R* paired hydrogen of fructose 6-phosphate.[32] 1-Tritioglucose 6-phosphate was isomerized to 1-tritiofructose 6-phosphate, and the biosynthetic fructose was oxidized with periodate to yield a tritiated glycolate sample derived from C-1 and C-2 of the fructose. The glycolate was purified and then oxidized to glyoxylate with glycolic acid oxidase; the tritium label was retained in the glyoxylate product. Since glycolic acid oxidase stereospecifically removes the *pro-R* hydrogen of glycolic acid (cf., footnote, p. 246), the tritiated glycolate isolated in this experiment must have been (*S*)-2-tritioglycolate. This, in turn, required that the original biosynthetic fructose possess the *S* configuration at the tritiated C-1 position (Figure VI-39). The hydrogen incorporated during the glucose

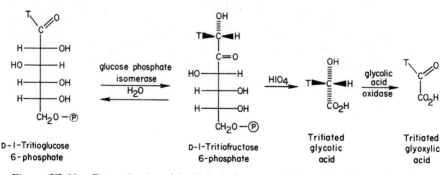

Figure VI-39. Determination of the biological stereospecificity of the glucose phosphate isomerase reaction.

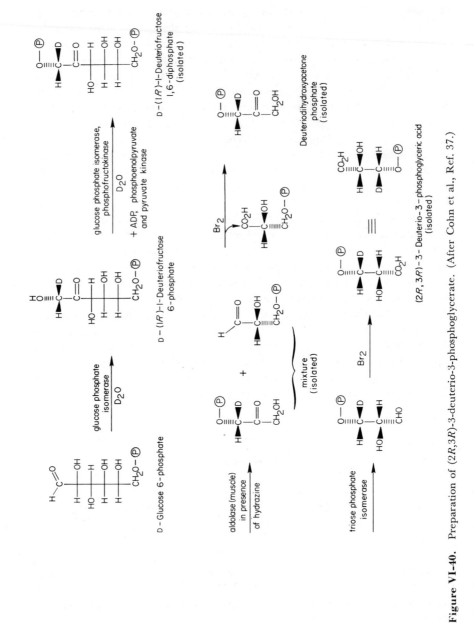

Figure VI-40. Preparation of (2R,3R)-3-deuterio-3-phosphoglycerate. (After Cohn et al., Ref. 37.)

phosphate isomerase-catalyzed conversion of 1-tritioglucose 6-phosphate must therefore have been incorporated into the *pro-R* position (Figure VI-39). In conjunction with Topper's results, this also established that mannose phosphate isomerase possesses a biological stereospecificity for the *pro-S* paired hydrogen at the C-1 position of fructose 6-phosphate.

The established stereospecificity of the glucose phosphate isomerase reaction has permitted the production of labeled biological intermediates that have been used to establish the stereospecificity of other important enzymatic processes. For example, the series of enzymatic reactions indicated in Figure VI-40 was used to convert the D-$(1R)$-1-deuteriofructose 6-phosphate obtained by incubating glucose 6-phosphate in D_2O with glucose phosphate isomerase into $(2R,3R)$-3-deuterio-3-phosphoglycerate [D-$(3R)$-3-deuterio-3-phosphoglycerate].[37] The isolated $(2R,3R)$-3-deuterio-3-phosphoglycerate was then incubated with phosphoglyceromutase (2,3-Diphospho-D-glycerate:2-phospho-D-glycerate phosphotransferase) and enolase (2-Phospho-D-glycerate hydro-lyase) to effect a final conversion to 3-deuteriophosphoenolpyruvate. As illustrated in Figure VI-41, the $(2R,3R)$-3-deuterio-3-phosphoglycerate substrate was initially converted to $(2R,3R)$-3-deuterio-2-phosphoglycerate by the mutase present in the incubation medium; then it was dehydrated to 3-deuteriophosphoenolpyruvate by the enolase present.

Figure VI-41 indicates that the stereochemistry of the deuterium labeling in the final 3-deuteriophosphoenolpyruvate product is dependent on the stereospecificity of the enolase reaction. The stereospecific *syn* elimination of water from a $(2R,3R)$-3-deuterio-2-phosphoglycerate will lead to a phosphoenolpyruvate with deuterium *cis* to the esterifying phosphate (Z isomer, see footnote to p. 148), whereas the stereospecific *anti* elimination of water from this intermediate will lead to a phosphoenolpyruvate with deuterium *trans* to the phosphate group (E isomer). The nmr spectra in Figure VI-42 present the vinyl hydrogen absorption of unlabeled phosphoenolpyruvate and the vinyl hydrogen absorption of the 3-deuteriophosphoenolpyruvate sample isolated by Cohn et al. after the sequence of reactions outlined in Figures VI-40 and VI-41. Clearly the vinyl hydrogen absorption centered at 5.33 δ (ppm downfield from TMS) is almost totally absent in the biosynthetic 3-deuteriophosphoenolpyruvate sample.

Because a deuterium nucleus does not absorb in the nmr, the spectra indicate that the vinyl hydrogen nucleus responsible for the 5.33 δ absorption in phosphoenolpyruvate has been stereospecifically replaced by a deuterium nucleus in the biosynthetic 3-deuteriophosphoenolpyruvate sample.* By carefully analyzing the vinyl nmr absorption of reference compounds, in

---

* The change in shape of the upfield absorption (5.15 δ) from two sets of overlapping doublets to a single doublet is also due to the deuterium substitution. The remaining vinyl hydrogen, now being *geminal* to a deuterium nucleus, is split only by the ^{31}P nucleus.

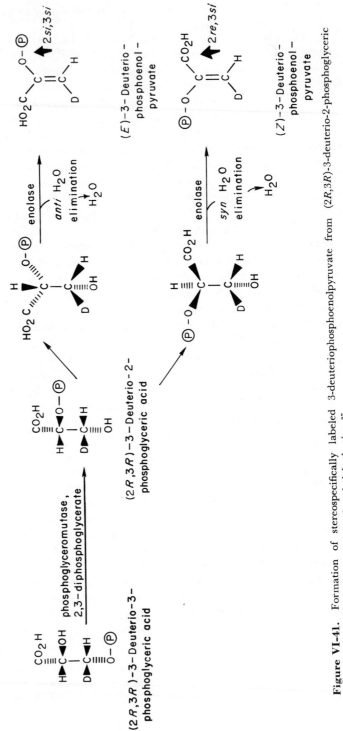

Figure VI-41. Formation of stereospecifically labeled 3-deuteriophosphoenolpyruvate from (2R,3R)-3-deuterio-2-phosphoglyceric acid via a stereospecific enolase-catalyzed dehydration.[37]

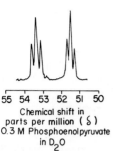

55 54 53 52 5I 50
Chemical shift in
parts per million (δ)
0.3 M Phosphoenolpyruvate
in D₂O

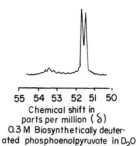

55 54 53 52 5I 50
Chemical shift in
parts per million (δ)
0.3 M Biosynthetically deuterated phosphoenolpyruvate in D₂O

Figure VI-42. Vinyl nmr absorption of phosphoenolpyruvate and biosynthetic 3-deuteriophosphoenolpyruvate samples.[37]

particular phosphoenolpyruvate-1-^{13}C, Cohn et al. were able to determine that the lower field vinyl hydrogen absorption of phosphoenolpyruvate (5.33 δ) was attributable to the vinyl hydrogen *trans* to the esterifying phosphate group. Since this vinyl hydrogen was replaced by deuterium in the biosynthetic 3-deuteriophosphoenolpyruvate sample, the enolase enzyme must stereospecifically catalyze an *anti* dehydration of 2-phosphoglycerate to yield (*E*)-3-deuteriophosphoenolpyruvate from the (3*R*)-3-deuterio substrate (Figure VI-41).

Considering the reverse reaction, the hydration of phosphoenolpyruvate to 2-phosphoglycerate catalyzed by enolase, the experimental results also establish that hydroxide addition must occur stereospecifically to C-3 on the 2*re*,3*re* face of the (*E*) deuterated phosphoenolpyruvate sample (*re,re* face of unlabeled phosphoenolpyruvate), whereas proton addition must occur stereospecifically to C-2 on the 2*si*,3*si* face (*si,si* face of unlabeled phosphoenolpyruvate). If this were *not* the case, deuterium would also have been found *cis* to the phosphate group in the 3-deuteriophosphoenolpyruvate sample isolated after the 1-hr incubation with enolase, *despite* the stereospecific *anti* dehydration process (Figure VI-41).

The established stereospecificities of the glucose phosphate isomerase and enolase reactions have been effectively utilized to prepare the labeled phosphoenolpyruvate molecules pictured in Figure VI-43. As described elsewhere in this book, these stereospecifically labeled phosphoenolpyruvates were used as substrates in the elegant series of experiments, carried out by Rose and his colleagues, which established the stereospecificities of carboxylations of phosphoenolpyruvate (cf. Chapter V, pp. 146–149), as well as the stereospecificities of methyl–methylene interconversions of phosphoenolpyruvate and malate (cf. pp. 234–239). It will be extremely instructive for the reader to determine the glucose phosphate isomerase and enolase incubation conditions required for the biosynthesis of the stereospecifically labeled

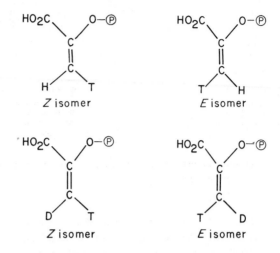

Figure VI-43. Labeled phosphoenolpyruvates prepared through the use of the biological stereospecificities of the glucose phosphate isomerase and enolase reactions.

phosphoenolpyruvate molecules in Figure VI-43 (cf. Figures VI-40 and VI-41).

Experiments involving the enzymatic isomerizations of isotopically labeled ketose and aldose phosphates have also provided information regarding the mechanisms of the ketol-isomerase reactions. Although incubation of a saccharide phosphate substrate with a ketol-isomerase under equilibrating conditions generally results in the incorporation of one hydrogen atom from the aqueous solvent into the ketose phosphate (at C-1) and one hydrogen atom from the solvent into the aldose phosphate (at C-2), experiments carried out under nonequilibrating conditions have demonstrated that a significant amount of intramolecular hydrogen transfer may occur during the ketol-isomerase reactions. Thus Rose and O'Connell studied the glucose phosphate isomerase-catalyzed conversion of biosynthetic $(1R)$-D-1-tritio-fructose 6-phosphate to D-glucose 6-phosphate under *nonequilibrating conditions** and found that some D-2-tritioglucose 6-phosphate was formed by the *intramolecular transfer of the tritium label.*[38] In the case of the isomerase enzyme from human red blood cells, the ratio of the tritium label lost from the $(1R)$-D-1-tritiofructose substrate into the solvent to the tritium label transferred to the C-2 atom of the glucose product was 0.97. With isomerase

* The reactions were carried out with rate-limiting amounts of isomerase in the presence of excess glucose 6-phosphate dehydrogenase (D-Glucose-6-phosphate:NADP oxidoreductase). In this way the glucose 6-phosphate formed was immediately converted to 6-phosphogluconate and was thereby removed from the isomerase-catalyzed equilibrium.

enzyme from rabbit muscle, the ratio of tritium lost to tritium transferred was 0.78, and with isomerase enzyme from yeast, this ratio was found to be 1.75. Simon and Medina found an exchange-to-transfer ratio of about 16 in the ketol-isomerase reaction catalyzed by mannose phosphate isomerase from yeast.[39] In the case of D-xylose isomerase (D-Xylose ketol-isomerase) from *Lactobacillus brevis*, which catalyzes the interconversion of D-xylose and D-xylulose, Rose et al. have reported that no substrate hydrogens were exchanged with the solvent protons *even under equilibrating conditions.*[40] With this particular enzyme, therefore, intramolecular transfer totally predominates. At the other extreme, Rieder and Rose failed to observe any evidence of intramolecular transfer in the ketol-isomerase reaction of dihydroxyacetone phosphate catalyzed by triose phosphate isomerase (D-Glyceraldehyde-3-phosphate ketol-isomerase).[41] Clearly the extent of intramolecular transfer involved in the ketol-isomerase reactions is a function of the particular enzyme studied. In addition, Rose and O'Connell found that the exchange-to-transfer ratio was markedly influenced by the incubation temperature.[38] With rabbit muscle glucose phosphate isomerase, the exchange-to-transfer ratio was 0.205 at an incubation temperature of 0° and increased to 3.17 at 60°. Thus lower incubation temperatures favored the intramolecular transfer of the $1R$ tritium label of fructose 6-phosphate to the C-2 position of glucose 6-phosphate; higher incubation temperatures, on the other hand, enhanced the exchange of the fructose $1R$ tritium label with the aqueous solvent (about 75% lost by exchange at 60°).

It is generally accepted that the ketol-isomerase reactions proceed via enzyme-bound enediol intermediates.* In the case of fructose 6-phosphate, for example, the stereospecific removal of the *pro-R* hydrogen at C-1 by a basic group on the enzyme and the transfer of a proton from an acidic group on the enzyme to the ketone carbonyl would yield an enediol intermediate. The isomerization to glucose 6-phosphate could then be completed by addition of the proton from the protonated basic group on the enzyme to the C-2 carbon and abstraction of a proton from the C-1 hydroxyl function. This process is illustrated in Figure VI-44. The catalyzed conversion of glucose 6-phosphate to fructose 6-phosphate would involve precisely the reverse sequence of enzymatically catalyzed proton abstractions and shifts.

As indicated in Figure VI-44, the proton on the basic catalytic group of the enzyme would be expected to undergo exchange with protons of the aqueous environment. This protonated-enzyme-H_2O exchange process would lead to either the stereospecific loss of label from the hexose substrates to the solvent, or the stereospecific incorporation of label from the solvent

---

* For a recent discussion of the possible mechanism of the glucose phosphate isomerase reaction, see Ref. 42.

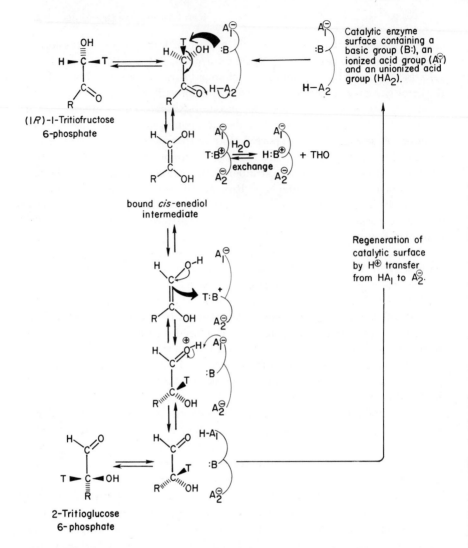

Figure VI-44. Schematic representation of the glucose phosphate isomerase reaction via a *cis*-enediol intermediate with solvent exchange and intramolecular transfer processes indicated.

into the hexose products. These exchange reactions should be facile, but they need not be instantaneous. The rate of this exchange reaction will depend on the accessibility of this region of the catalytic site of the enzyme to molecules of solvent water. Various factors, therefore, including the "tightness" of the active enzymatic conformation and the rapidity of reprotonation of the bound enediol intermediate will determine the observed ratio of intramolecular proton transfer to the loss of label from the hexoses to solvent for a particular isomerase reaction. In this way a single fundamental ketol-isomerization mechanism such as outlined in Figure VI-44 can explain the variable transfer to exchange ratios observed experimentally.

Since the postulated enediol intermediates possess biologically differentiable *pro*-chiral faces, the *observed stereospecificities* of the hydrogen exchange processes *could potentially occur via free enediol intermediates*. A free enediol, however, would be expected to undergo noncatalyzed tautomerizations to regenerate the ketose–aldose structures at an appreciable rate. Such non-catalyzed tautomerizations of free enediol intermediates would result in the eventual incorporation of two protons from the solvent into the ketose phosphates. *The observed stereospecific exchange of only a single hydrogen* of the ketose phosphate in ketol-isomerase-catalyzed reactions under equilibrating conditions thus requires the enzyme-bound enediol intermediate of Figure VI-44 (cf. Ref. 41).

The absolute biological stereospecificity of the transfer or exchange of the paired hydrogens at C-1 of the ketose will depend on the required geometrical relationship between the ketose phosphate and the enzymatic basic group. The biological stereospecificity of reaction toward the C-2 chiral group of the aldose phosphate will be determined by the geometrical relationship between the bound enediol intermediate and the enzymatic group which adds a proton to the C-2 carbon center. At present all ketol-isomerase reactions of known absolute stereospecificities obey the following correlation, originally noted by Rose and O'Connell.[32] Ketol-isomerases that are known to stereospecifically remove the *pro-R* hydrogen at C-1 of the ketose phosphates lead to aldose phosphates with the R (D) absolute configuration at C-2 (the glucose phosphate, ribose phosphate, triose phosphate, D-xylose, and L-arabinose ketol-isomerases). On the other hand, the ketol-isomerases known to stereospecifically remove the *pro-S* paired hydrogen at C-1 of the ketose phosphates lead to aldose phosphates with the S (L) absolute configuration at C-2 (the mannose phosphate and L-fucose ketol-isomerases).

The postulated enzyme-bound enediol intermediates of the ketol-isomerase reactions could possess either the *cis* stereochemistry pictured in Figure VI-44 or the *trans* stereochemistry of Figure VI-45. If an enzyme-bound *cis*-enediol intermediate is involved in *both* the glucose phosphate and mannose phosphate isomerase reactions, then the established biological

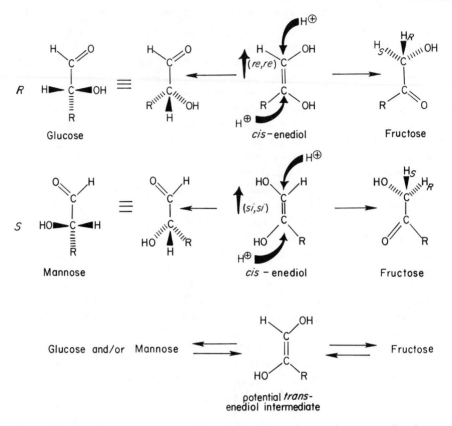

Figure VI-45. Required stereospecificity of glucose phosphate and mannose phosphate isomerase reactions with postulated *cis*-enediol intermediates.

stereospecificities require that both proton abstraction and addition occur on one face of the *cis*-enediol system in the glucose phosphate isomerase-catalyzed process and on the opposite face in the mannose phosphate isomerase-catalyzed process. As shown in Figure VI-45, the established stereospecificity of the glucose phosphate isomerase reaction necessitates proton abstractions and additions at C-1 and C-2 of the postulated *cis*-enediol intermediate exclusively from the *re,re* face of the enediol plane. The biological stereospecificity of the mannose phosphate isomerase reaction, on the other hand, would necessitate proton abstractions and additions at C-1 and C-2 of the bound *cis*-enediol intermediate exclusively from the *si,si* face of the enediol plane (Figure VI-45). If a bound *trans*-enediol intermediate were involved in either or both of these isomerizations, then a different set of

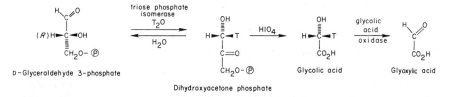

Figure VI-46. Stereospecificity of the triose phosphate isomerase reaction.

stereospecific proton abstractions and additions would be required. The reader should determine the kind of stereospecific additions to a *trans*-enediol intermediate that would be required in the glucose phosphate and mannose phosphate isomerase reactions.

If the foregoing analyses have been correctly carried out, an interesting contrast in the requirements of *cis*-enediol and *trans*-enediol intermediates in ketol-isomerase processes will become apparent. Let us make the reasonable assumption that the intramolecular proton transfers observed in ketol-isomerase reactions are indicative of mechanisms involving proton additions and abstractions on *one side of the enediol intermediates*. In that case, the correlation originally noted by Rose and O'Connell,[32] namely, *pro-R* hydrogen activation → (R) C-2 chirality and *pro-S* hydrogen activation → (S) C-2 chirality, is experimental evidence for *cis*-enediol intermediates in the ketol-isomerase reactions (cf. Ref. 40).

The triose phosphate isomerase reaction, which interconverts dihydroxyacetone phosphate and D-glyceraldehyde-3-phosphate, is another common ketol-isomerase. It has been determined that this enzyme stereospecifically removes* the *pro-R* paired hydrogen of the ketose phosphate[34] (cf. Figure VI-46). Clearly the enzyme must also stereospecifically convert the ketone carbonyl into an R (D) secondary alcohol group in order to yield D-glyceraldehyde 3-phosphate (Figure VI-46). Dihydroxyacetone phosphate lacks chiral centers, and thus the paired hydrogen atoms of this substrate which are differentiated in the triose phosphate isomerase reaction are enantiotopic rather than diastereotopic. This instance of biological stereospecificity between paired methylene hydrogens therefore differs somewhat from the stereospecificity of the ketol-isomerases discussed earlier. In fact, however, the biological stereospecificity of this particular ketol-isomerase for paired enantiotopic hydrogens merely emphasizes that enzymatic differentiation between both enantiotopic and diastereotopic nonequivalent paired groups is based on the fundamental geometrical relationships generated in the substrate-enzyme complex.

_____

* No intramolecular transfer could be detected.[41]

BIOLOGICAL STEREOSPECIFICITY OF THE ALDOLASE REACTION

Dihydroxyacetone phosphate is also involved in the vital aldolase (Fructose-1,6-diphosphate D-glyceraldehyde-3-phosphate-lyase) reaction of intermediary metabolism. Figure VI-47 shows that the aldol reaction catalyzed by this enzymatic system (the enzyme catalyzes both the condensation and the reverse cleavage reaction), must result in the loss of one of the paired enantiotopic hydrogens of dihydroxyacetone phosphate and its replacement by a new carbon–carbon bond. Therefore it is anticipated that aldolase, like triose phosphate isomerase, will manifest biological stereospecificity toward one of the nonequivalent paired hydrogens of the hydroxymethyl group of dihydroxyacetone phosphate. As in the studies of the ketol-isomerase stereospecificities, studies of the biological stereospecificity of the aldolase reaction have yielded information of mechanistic importance.

In the absence of glyceraldehyde 3-phosphate, both muscle aldolase (type I aldolase) and yeast aldolase (type II aldolase) have been found to catalyze the exchange of one of the enantiotopic paired hydrogens of dihydroxyacetone phosphate with the protons of the aqueous solvent.[43] The sequence outlined in Figure VI-48 was used by Rose to establish the stereospecificity of this catalyzed exchange.[34]

Dihydroxyacetone phosphate was labeled with tritium by incubation with aldolase in the presence of tritiated water. The tritiated dihydroxy-

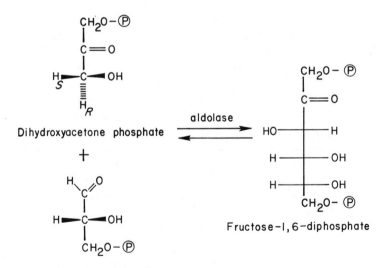

Figure VI-47. Aldol condensation and the reverse cleavage catalyzed by aldolase.

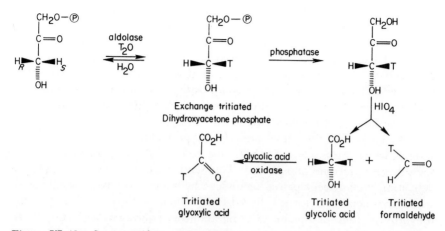

Figure VI-48. Stereospecificity of the aldolase-catalyzed exchange reaction with dihydroxyacetone phosphate.

acetone phosphate was then treated with phosphatase and the resulting dihydroxyacetone was oxidized to glycolate and formaldehyde with periodate. The glycolate was then purified and oxidized to glyoxylate with glycolic acid oxidase. The final glyoxylate had the same tritium specific activity as the glycolate sample and about one-half the specific activity of the original enzymatically tritiated dihydroxyacetone. This is the result that would be expected, given a stereospecific aldolase-catalyzed exchange.* Furthermore, the established stereospecificity of the glycolic acid oxidase reaction proved that the *pro-S* hydroxymethyl hydrogen of the dihydroxyacetone phosphate was stereospecifically replaced by tritium (Figure VI-48).

Aldolase and triose phosphate isomerase therefore display opposite absolute biological stereospecificities toward the enantiotopic paired hydrogens at C-1 of dihydroxyacetone 3-phosphate. As would be predicted, when the (S)-1-tritiodihydroxyacetone 3-phosphate prepared by aldolase-catalyzed

---

* It should be remembered that the rotational symmetry possessed by the dihydroxyacetone molecule results in superimposable hydroxymethyl groups. These groups are thus geometrically equivalent and not subject to differentiation (cf. p. 162). The periodate oxidation of dihydroxyacetone will therefore lead to random production of formaldehyde from each of two equivalent hydroxymethyl groups. The glycolate will likewise be produced from the central carbon plus *either* of the two equivalent hydroxymethyl groups. Only the nonesterified hydroxymethyl group of the original dihydroxyacetone phosphate molecule of this experiment bears a tritium label. The C-2 (hydroxymethyl group) of the glycolate will therefore be randomly derived from both the labeled and unlabeled terminal positions, and the glycolate specific activity should be only half that of the original dihydroxyacetone phosphate.

exchange was equilibrated in H_2O with triose phosphate isomerase, the tritium label was not lost from the trioses.[43]

The available experimental evidence indicates that the aldolase-catalyzed exchange reaction observed in the absence of glyceraldehyde 3-phosphate is not an artifactual side reaction but is intimately connected with the mechanism of the aldolase condensation. The observed stereospecific exchange reaction, therefore, provides some insight into the probable enzymatic mechanism of the condensation.

We know that the active site of the type I aldolases contains a lysine residue, which forms a Schiff base (imino bond) with the ketone carbonyl of the dihydroxyacetone phosphate.[44] The Schiff base formation is viewed as activating the adjacent paired methylene hydrogens. (In type II aldolases, the bound-metal-cofactor is believed to coordinate with the carbonyl of the dihydroxyacetone phosphate to bring about an equivalent activation.[45]) A basic group within the active site of the enzyme is then postulated to stereospecifically remove one of the activated methylene hydrogens to yield a bound enamine intermediate and a protonated enzymatic basic group. As in the previously described ketol-isomerase system, the proton carried on the enzymatic basic group would be expected to undergo exchange with protons present in the aqueous solvent. Such exchange, followed by reprotonation of the enamine intermediate and hydrolysis of the Schiff base linkage, serves to explain the experimentally observed stereospecific hydrogen exchange catalyzed by the aldolase enzyme.[34,43]

In the presence of glyceraldehyde 3-phosphate or other suitable aldehydes, however, a nucleophilic attack by the enamine on the aldehydic carbonyl can occur instead of a nucleophilic attack on the proton attached to the enzymatic basic group. This attack on the aldehydic carbonyl leads to the formation of a new carbon–carbon bond instead of the simple reprotonation process. Hydrolysis of the Schiff base linkage then yields the overall aldolase condensation product. This mechanism, illustrated schematically in Figure VI-49, will explain how stereospecific exchange of one of the hydrogens of dihydroxyacetone phosphate with solvent protons can occur in the absence of the aldehyde substrates, whereas in the presence of glyceraldehyde 3-phosphate, no incorporation of solvent hydrogens into the fructose 1,6-diphosphate product is detected.

Although the aldolase enzyme is specific for the dihydroxyacetone phosphate moiety, a variety of aldehydes will function as active substrates. In each case, however, the condensing aldehyde carbonyl is converted into a secondary alcohol group with the R (D) absolute configuration (e.g., C-4 of fructose 1,6-diphosphate illustrated in Figure VI-47). This requires that the nucleophilic attack of the enamine intermediate occur stereospecifically on the si face of the aldehydic carbonyl. At the same time, aldolase stereo-

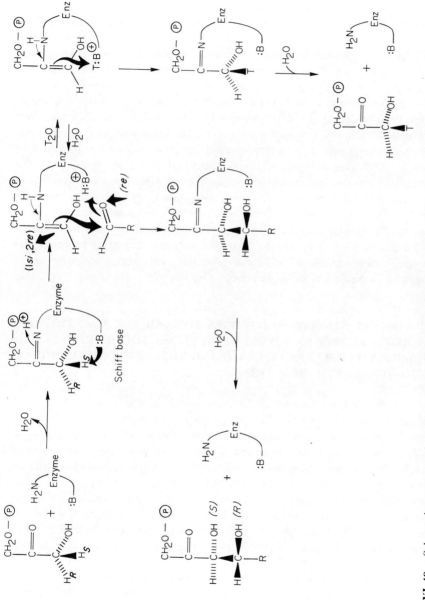

Figure VI-49. Schematic representation of the aldolase-catalyzed processes, indicating observed stereospecificities.

specifically converts the condensing *pro*-chiral hydroxymethyl group of the dihydroxyacetone phosphate moiety into a secondary alcohol group in the product with S (L) absolute chirality (e.g., C-3 of the fructose 1,6-diphosphate product illustrated in Figure VI-47). Although the eneamine intermediate may possess the *cis* arrangement of amino and hydroxyl functions illustrated in Figure VI-49, the eneamine intermediate could possess the alternative *trans* arrangement of these groups. The stereospecific removal of the *pro-S* hydrogen from dihydroxyacetone phosphate and the stereospecific formation of an S chiral center at C-3 in the product, however, requires the reaction to occur with retention at this condensing hydroxymethyl group. That is, both the proton removal and exchange process, and the carbon–carbon bond-forming process with the condensing aldehyde group must occur on the same face of the eneamine plane. This is illustrated in Figure VI-49 with the potential *cis*-eneamine intermediate. In the *cis*-eneamine case, the reprotonation and the carbon–carbon bond formation must both occur from the $1si,2re$ face of the *cis*-eneamine intermediate to produce the observed stereospecific H_S exchange in the dihydroxyacetone phosphate and the observed stereospecific (S) C-3 hydroxyl production in the condensation product. The required stereospecificity of these processes with the alternative *trans*-eneamine is again left as an exercise.

BIOLOGICAL STEREOSPECIFICITIES TOWARD THE ENANTIOTOPIC PAIRED α-HYDROGEN ATOMS OF GLYCINE: THE SERINE HYDROXYMETHYLTRANSFERASE AND THE δ-AMINOLEVULINIC ACID SYNTHETASE REACTIONS

In contrast to the rest of the α-amino acids, the α position (C-2) of glycine is a *pro*-chiral center. The paired methylene hydrogen atoms attached to this α-*pro*-chiral center are enantiotopic. Investigations carried out by Akhtar and Jordan have demonstrated that the enantiotopic α-hydrogens of glycine are biochemically nonequivalent in the serine hydroxymethyltransferase reaction[46] and in the δ-aminolevulinic acid synthetase reaction.[47] The biological stereospecificity between the paired enantiotopic glycine hydrogens detected in the δ-aminolevulinic synthetase reaction is an especially interesting example. Here the detected biological stereospecificity, in and of itself, permits us to differentiate between two possible enzymatic reaction mechanisms.

The formation of L-serine from glycine in the reaction catalyzed by serine hydroxymethyltransferase (L-Serine:tetrahydrofolate 5,10-hydroxymethyltransferase) emphasizes the *pro*-chiral nature of the C-2 position of glycine (Figure VI-50). Akhtar and Jordan found that when glycine was

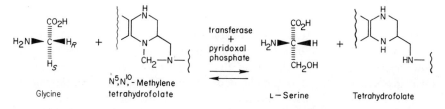

Figure VI-50. Formation of L-serine from glycine by serine hydroxymethyltransferase.

incubated with serine hydroxymethyltransferase in the presence of pyridoxal phosphate and tetrahydrofolic acid, *but in the absence of formaldehyde,* one of the methylene hydrogens of glycine underwent rapid stereospecific exchange with the protons of the solvent.[47] Using these experimental conditions, Akhtar and Jordan prepared stereoisomeric samples of 2-tritioglycine by incubating the commercially available, nonselectively labeled 2-tritioglycine [(*RS*)-2-tritioglycine] in H_2O and by incubating unlabeled glycine in T_2O. Reincubation of each of the resulting 2-tritioglycine samples in H_2O with serine hydroxymethyltransferase under these special conditions established that the observed exchange process was stereospecific. As anticipated, the 2-tritioglycine sample prepared by incubating commercially available 2-tritioglycine in H_2O lost only a small amount of the tritium label upon re-incubation in H_2O. In contrast, the 2-tritioglycine sample prepared by incubating unlabeled glycine in T_2O lost over 85% of the tritium label when reincubated in H_2O.

The absolute configuration of these 2-tritioglycine samples, and thus the absolute stereospecificity of the serine hydroxymethyltransferase-catalyzed exchange, was established via the procedure outlined in Figure VI-51.[46]

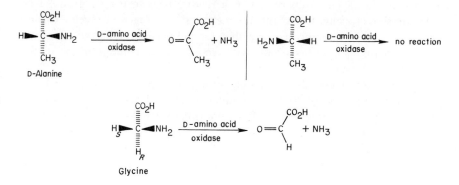

Figure VI-51. Oxidation of glycine by D-amino acid oxidase.

As discussed in Chapter IV, D-amino acid oxidase has stereospecificity for amino acid substrates of the D (or S) absolute configuration. D-Alanine is converted into pyruvate by this enzyme, whereas L-alanine is inert in the enzymatic reaction. Purified preparations of D-amino acid oxidase will, however, catalyze the oxidation of the C-2 *pro*-chiral center of glycine to form glyoxylate. Since, as indicated in Figure VI-51, the oxidation of D-alanine necessarily involves the removal of an α-hydrogen whose absolute configuration corresponds to that of H_S at the *pro*-chiral center of a glycine substrate, it can be inferred that the oxidation of glycine to glyoxylate catalyzed by D-amino acid oxidase will involve the stereospecific removal of the H_S paired methylene hydrogen. (Analogous reasoning was previously used to tentatively establish the stereospecificity of the glycolic acid oxidase system; cf. Figure VI-37.) When the 2-tritioglycine sample obtained by equilibrating unlabeled glycine in tritiated water was oxidized with D-amino acid oxidase, the glyoxylate product retained only 16% of the original tritium label. The 2-tritioglycine sample obtained by equilibrating (RS)-2-tritioglycine in H_2O, however, yielded a glyoxylate product that retained 83% of the original tritium label. Therefore, we may infer that the former glycine sample was (S)-2-tritioglycine and the latter sample was the (R)-2-tritioglycine enantiomer. This result established that the incubation of glycine with serine hydroxymethyltransferase under Akhtar and Jordan's special conditions causes the stereospecific equilibration of the H_S paired hydrogen with protons of the solvent.

Akhtar and Jordan next incubated the separate (R) and (S)-2-tritioglycine samples with serine hydroxymethyltransferase and the necessary cofactors *in the presence of formaldehyde* to produce L-serine.[46] Analysis of the biosynthetic L-serine samples revealed that the (S)-2-tritioglycine sample yielded unlabeled L-serine, whereas the (R)-2-tritioglycine sample yielded an L-2-tritioserine sample containing the total tritium label originally present in the glycine substrate. It can readily be seen from Figure VI-50 that the stereospecific loss of the H_S methylene hydrogen of glycine during the formation of L-serine requires that the conversion catalyzed by serine hydroxymethyltransferase occur with retention of configuration at the α-carbon of glycine.

The enantiomeric 2-tritioglycine substrates were also used by Akhtar and Jordan to establish that δ-aminolevulinic acid synthetase (δ-Aminolevulinic acid dehydratase) manifests biological stereospecificity toward the paired methylene hydrogens of glycine.[47] In the presence of magnesium ion and pyridoxal phosphate, δ-aminolevulinic acid synthetase catalyzes the condensation of glycine and succinyl CoA to form δ-aminolevulinate as outlined in Figure VI-52. The δ-aminolevulinate is then utilized as a key intermediate in the biosyntheses of porphyrins, chlorophylls, and corrinoids.

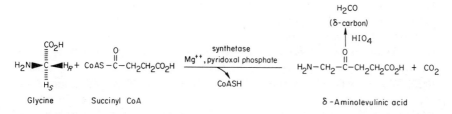

Figure VI-52.   Biosynthesis of δ-aminolevulinate by δ-aminolevulinic acid synthetase.

Akhtar and Jordan briefly incubated the commercially available, nonspecifically labeled 2-tritioglycine and each of the two stereospecifically labeled 2-tritioglycine samples with δ-aminolevulinic acid synthetase.* The δ-methylene group of the δ-aminolevulinate product formed in each incubation was converted to formaldehyde by periodate oxidation and isolated as the formaldehyde-dimedone adduct (Figure VI-52). This δ-methylene group of the δ-aminolevulinate product is derived biosynthetically from the methylene group of glycine. By adding a small amount of glycine-2-^{14}C to each 2-tritioglycine sample and determining the ^{3}H/^{14}C labeling ratios in the original glycine and in the final formaldehyde-dimedone adducts, Akhtar and Jordan were able to determine the amount of tritium label lost from the methylene position of each glycine sample during the enzyme catalyzed formation of δ-aminolevulinate. The experimental results established that about 57% of the tritium label in the (RS)-2-tritioglycine sample was lost during the conversion of the methylene group of glycine into the δ-methylene group of δ-aminolevulinate. Furthermore, 25% of the tritium label in the (S)-2-tritioglycine sample and 97% of the tritium label in the (R)-2-tritioglycine sample were lost in the conversion catalyzed by δ-aminolevulinic acid synthetase. These results establish that the δ-aminolevulinic acid synthetase reaction stereospecifically eliminates the H_R paired methylene hydrogen of glycine during the conversion of this group into the δ-methylene group of δ-aminolevulinate. The glycine α-hydrogen stereospecifically exchanged (or eliminated) in the serine hydroxymethyltransferase reaction therefore possesses an absolute configuration opposite to that of the glycine α-hydrogen stereospecifically lost in the δ-aminolevulinic acid synthetase reaction.

This demonstration of biological stereospecificity toward the paired methylene hydrogens of glycine in the δ-aminolevulinic acid synthetase reaction provides evidence of a probable reaction mechanism. Initially two fundamentally different condensation processes seem possible. As illustrated in

---

* The brief incubation was designed to reduce the extent of the exchange of the δ-methylene hydrogens that was known to occur in the final δ-aminolevulinate product.

Figure VI-52, the overall condensation process results in the loss of the carboxyl group of the glycine moiety. It is possible, therefore, that a glycine–pyridoxal phosphate complex could undergo an initial decarboxylation and the resulting stabilized anion could then participate in a nucleophilic attack on the thioester function of succinyl CoA to form the new carbon–carbon bond (Figure VI-53A). Alternatively, a glycine–pyridoxal phosphate complex might lose a methylene hydrogen from the glycine moiety to yield a stabilized carbanion, which could then participate in a nucleophilic attack

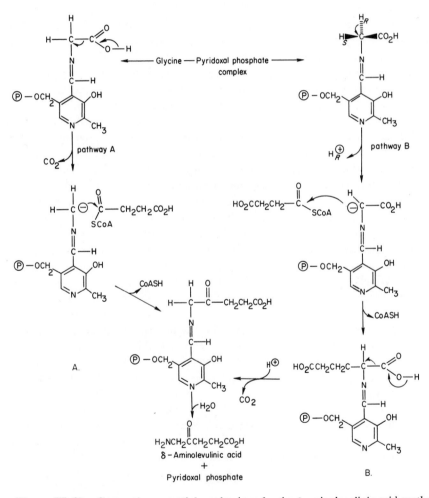

Figure VI-53. Contrasting potential mechanisms for the δ-aminolevulinic acid synthetase reaction.

on the thioester function of succinyl CoA to form the new carbon–carbon bond. In the latter mechanism, the glycine moiety would undergo decarboxylation *after* the formation of a new carbon–carbon bond linking the glycine and succinyl CoA moieties (Figure VI-53B). The demonstration that the δ-aminolevulinic acid synthetase reaction stereospecifically activates and exchanges one of the methylene hydrogens of the glycine moiety is more consistent with the second hypothesized mechanism (Figure VI-53B).*

BIOLOGICAL STEREOSPECIFICITIES IN REACTIONS INVOLVING COBALAMIN (B_{12}) COENZYMES: STEREOCHEMICAL EVIDENCE OF PARTICULAR REACTION PATHWAYS

Enzymatic reactions involving the cobalamin coenzymes (vitamin B_{12} coenzymes) have been selected as our final instructive examples of stereospecific biological processes. These reactions display many varied features of stereochemical interest. For example, the sequence indicated in Figure VI-54 forms a vital portion of the metabolism of propionic acid. In this sequence, a cobalamin-requiring enzyme catalyzes the interconversion of methylmalonyl CoA and succinyl CoA. The first stereochemically interesting feature of this reaction is the steric specificity of methylmalonyl CoA mutase (Methylmalonyl–CoA CoA–carbonylmutase) for the (R)-methylmalonyl CoA enantiomer. As indicated in Figure VI-54, however, a prior enzyme in this metabolic sequence, propionyl CoA carboxylase [Propionyl–CoA:carbondioxide ligase (ADP)], stereospecifically catalyzes the formation of the (S)-methylmalonyl CoA enantiomer. The overall metabolic sequence thus requires the presence of a racemase (Methylmalonyl–CoA racemase) to interconvert the (R) and (S)-methylmalonyl CoA enantiomers, thereby coupling the carboxylase and mutase-catalyzed steps.

The reaction catalyzed by the cobalamin-requiring methylmalonyl CoA mutase appears in greater detail in Figure VI-55. It is clear that this enzymatic reaction involves an interconversion of methyl and methylene groups.

The three different classes of enzymatically catalyzed methyl–methylene interconversions indicated in Figure VI-56 have now been presented in this chapter. The first class of enzyme-catalyzed methyl–methylene interconversions includes condensations involving acetyl CoA, such as the citrate synthetase reactions. The recent investigations[9–13] with chiral acetates, which elucidated the stereochemistry of several examples of this first class of methyl–methylene interconversions, were discussed earlier (pp. 194–211). The iso-

---

* What additional mechanistic information could be obtained if the configuration of the final δ-methylene group of an appropriate δ-aminolevulinate product molecule were determined?

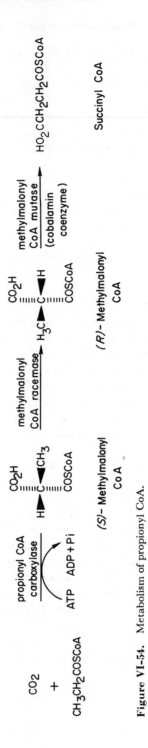

Figure VI-54. Metabolism of propionyl CoA.

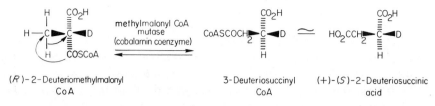

Figure VI-55. Determination of the stereospecificity of the methylmalonyl CoA mutase reaction.

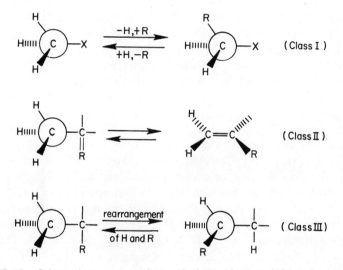

Figure VI-56. Schematic representations of three classes of biological methyl–methylene interconversions.

pentenyl pyrophosphate isomerase and the pyruvate kinase reactions are examples of the second class of enzymatic methyl–methylene interconversions. As previously stated (pp. 231–234), the isopentenyl pyrophosphate isomerase reaction is the only step of undetermined stereochemistry in the mevalonate to squalene biosynthetic sequence.* On the other hand, recent experiments by Rose have elucidated the stereospecificity of the pyruvate kinase reaction. The methylmalonyl CoA mutase reaction of Figure VI-55 involves a 1,2-shift of a methyl hydrogen and a simultaneous 1,2-shift of the acyl CoA group. This process interconverts a methyl group and a saturated methylene position in a fashion that is distinct from the first two types of enzymatic methyl–methylene interconversions. The mutase reactions of this type,

---

* Note added in proof: Clifford et al. have now established the stereospecificity of this step, cf. Ref. 71.

catalyzed by cobalamin-requiring enzymes, thus constitute a third class of enzymatic methyl–methylene interconversions.

The stereochemistries of the transformations at the methyl carbon centers in the methylmalonyl CoA mutase and the related methylaspartate mutase (L-*threo*-3-Methylaspartate carboxy-aminomethylmutase, cf. Figure VI-57) reactions have not yet been determined. As in the previously discussed case of the isopentenyl pyrophosphate isomerase reaction, however, the recently developed techniques of preparing and analyzing chiral acetate molecules now make it feasible to begin to investigate these stereochemical questions. The precise nature of the stereochemical changes that occur at the methyl functions in methyl–methylene interconversions of this third class will undoubtedly be established in the near future.

Although we still lack knowledge regarding the stereochemical fate of the methyl hydrogens in the methylmalonyl CoA mutase reaction, the absolute stereochemistry of the rearrangement at the adjacent center has been determined.[48] An enzyme extract from beef liver mitochondria, containing propionyl CoA carboxylase, methylmalonyl CoA racemase, and methylmalonyl CoA mutase activities, was used to convert propionyl CoA to succinyl CoA (cf. Figure VI-54). The enzymatic incubation was carried out in D_2O. Under these conditions, an exchange reaction catalyzed by the methylmalonyl CoA racemase resulted in the *in situ* formation of (R)-2-deuteriomethylmalonyl CoA, which was converted by the mutase reaction to 3-deuteriosuccinyl CoA.* The deuterated succinyl CoA was then hydrolyzed to 2-deuteriosuccinate and was found to be dextrorotatory. This monodeuterated succinate could therefore be assigned the S absolute configuration, based on the levorotatory (R)-2-deuteriosuccinate reference standard previously prepared by Cornforth and Popják (cf. Figure VI-18). As illustrated in Figure VI-55, the conversion of (R)-2-deuteriomethylmalonyl CoA to $(3S)$-3-deuteriosuccinyl CoA requires that the carbon–carbon bond at C-2 of the (R)-methylmalonyl CoA be replaced by a new carbon–hydrogen bond of the same absolute stereochemistry at C-3 in the succinyl CoA product. The methylmalonyl CoA mutase catalyzed rearrangement therefore proceeds with *retention of configuration* at C-2 of the original methylmalonyl CoA molecule.

The methylaspartate mutase reaction illustrated in Figure VI-57 is also catalyzed by an enzyme that requires a cobalamin coenzyme. Although the overall rearrangement catalyzed by this cobalamin-requiring mutase seems analogous to the methylmalonyl CoA mutase rearrangement, it has been established that the methylaspartate rearrangement proceeds with an inver-

---

* Other experiments, previously carried out,[49,50] indicated that no exchange with the solvent hydrogens occurred during the methylmalonyl CoA mutase reaction.

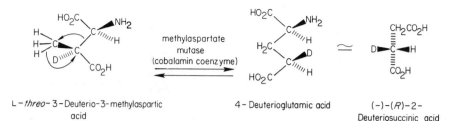

L-*threo*-3-Deuterio-3-methylaspartic acid   4-Deuterioglutamic acid   (−)-(R)-2-Deuteriosuccinic acid

Figure VI-57. Determination of the stereospecificity of the methylasparate mutase reaction.

sion of configuration at the analogous C-3 carbon of the original L-*threo*-3-methylaspartate.[51] The established stereospecificity of this process is indicated in Figure VI-57. L-*Threo*-3-deuterio-3-methylaspartate was rearranged with methylaspartate mutase,* and the 4-deuterioglutamate product was isolated and chemically oxidized to yield a 2-deuteriosuccinate molecule. This deuterated succinate was found to be levorotatory and was therefore (R)-2-deuteriosuccinate. This experimental result requires the methylaspartate mutase reaction to proceed with a net inversion of configuration at C-3 of the rearranging L-*threo*-3-deuterio-3-methylaspartate substrate (Figure VI-57).

The diol dehydrase reaction (Propanediol hydro-lyase) of Figure VI-58A is another process catalyzed by a cobalamin-requiring enzyme. This reaction could conceivably have involved dehydration to the enol and then tautomerization; however, we know that it proceeds via a substrate rearrangement, as indicated in Figure VI-58B. Although the diol dehydrase reaction obviously does not involve a methyl–methylene interconversion, Figure VI-58B emphasizes the similarity of this reaction to the mutase-catalyzed rearrangements outlined in Figures VI-55 and VI-57. The diol dehydrase enzyme is unusual in that both (R) and (S)-propane-1,2-diol serve as effective substrates. Despite this noteworthy lack of steric specificity between enantiomeric substrate molecules,† the enzyme manifests a high degree of biological stereospecificity in the remaining features of the reaction process. For example, independent experiments carried out by Zagalak et al.[54] and Rétey et al.[55] established that the dehydrase reaction proceeds with net

---

* Again it was known that no exchange of hydrogen with the water took place during the mutase-catalyzed rearrangement.[52]

† Actually, diol dehydrase does not totally lack a substrate steric specificity. The enzyme has been found to possess a slight steric specificity for the (R)-propane-1,2-diol substrate. The (R) enantiomer was found to be about 1.5 to 2 times as active as the (S) enantiomer. In addition, the (S) enantiomer slightly inhibited conversion of the more active (R)-propane-1,2-diol substrate.[53]

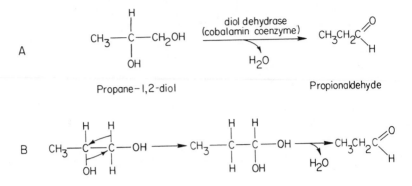

Figure VI-58. Propane-1,2-diol dehydrase reaction.

inversion of configuration at the C-2 carbon of the reacting propane-1,2-diol substrates. In this respect, therefore, the diol dehydrase and the methyl-aspartate mutase reactions are analogous and may be contrasted with the methylmalonyl CoA mutase reaction. The reason for these contrasting stereo-chemical results in different cobalamin-enzyme-catalyzed rearrangements is not yet known.

The methylene hydrogens at the C-1 position of the propane-1,2-diols are nonequivalent, diastereotopic paired substituents. It is anticipated, there-fore, that the diol dehydrase reaction will differentiate between these paired hydrogens in the rearrangement process. This aspect of the biological stereo-specificity associated with the diol dehydrase reaction was also investigated by Zagalak et al.[54] (R)-Lactaldehyde was reduced with (4'R)-NAD-²H and alcohol dehydrogenase. The established stereospecificity of alcohol dehydro-genase (cf. pp. 159–161) resulted in the production of (1R,2R)-1-deuterio-propane-1,2-diol from this reaction. This substrate was then incubated with diol dehydrase, the propionaldehyde product was isolated as a 2,4-dinitro-phenylhydrazone derivative, and the position of the deuterium in the propionaldehyde product was determined by nmr spectroscopy. It was found that the diol dehydrase-catalyzed rearrangement of (1R,2R)-1-deuterio-propane-1,2-diol yielded a 2-deuteriopropionaldehyde product (Figure VI-59). Since diol dehydrase reacts with both enantiomeric stereoisomers of the propane-1,2-diol substrate, the biological stereospecificity between the paired methylene hydrogens manifest in reactions of the (S)-propane-1,2-diol sub-strate was also investigated. (S)-Lactaldehyde was reduced with (4'R)-NAD-²H and alcohol dehydrogenase to yield a (1R,2S)-1-deuteriopropane-1,2-diol substrate. This specifically deuterated propanediol substrate was rearranged in the presence of diol dehydrase and the deuterated propion-

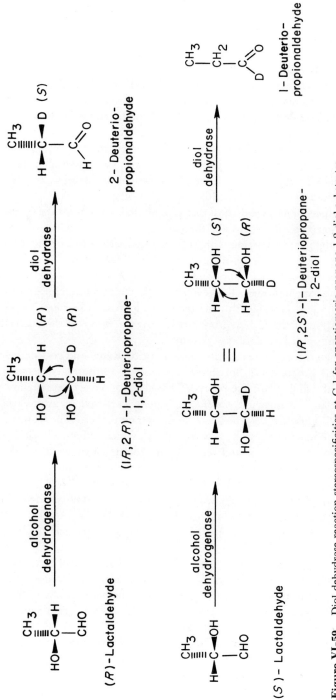

Figure VI-59. Diol dehydrase reaction stereospecificities at C-1 for enantiomeric propane-1,2-diol substrates.

275

aldehyde product analyzed as outlined previously. The nmr analysis established that the product of the (1R,2S)-1-deuteriopropane-1,2-diol rearrangement is 1-deuteriopropionaldehyde (Figure VI-59).

These experimental results of Zagalak et al. thus established that diol dehydrase manifests biological stereospecificity toward the paired diastereotopic methylene hydrogens of propane-1,2-diol, *despite* its lack of steric specificity toward the propanediol enantiomers. In the reaction of the (R)-propane-1,2-diol, the *pro-R* hydrogen at C-1 is stereospecifically rearranged; in the reaction of the (S)-propane-1,2-diol, the *pro-S* hydrogen at C-1 is stereospecifically rearranged. These stereochemical aspects of the diol dehydrase reaction are presented in Figure VI-59.

The experimental results just described make it apparent that, although biological stereospecificity toward the paired methylene hydrogens is evident in the diol dehydrase reaction with both enantiomeric forms of the substrate, the absolute stereospecificity of this enzymatic differentiation is opposite for the two propane-1,2-diol enantiomers. The projection drawings of Figure VI-59 emphasize, however, that even though the absolute chirality of that paired hydrogen rearranged in the two enantiomeric substrates is opposite, the orientation of the C-1 hydrogen rearranged relative to the rearranging hydroxyl group at C-2 remains constant.

An elegant set of experiments carried out by Rétey and co-workers with ^{18}O-enriched propane-1,2-diols have illuminated additional stereochemical features of the diol dehydrase reaction.[56] When acetoxyacetone is allowed to equilibrate in ^{18}O-enriched water, the ketone carbonyl of acetoxyacetone becomes ^{18}O-enriched as a result of an exchange process. The ^{18}O-enriched acetoxyacetone was then treated with LiAlH$_4$, which reductively cleaved the esterifying acetoxyl group and reduced the ketone carbonyl to yield (RS)-1,2-propanediol-2-^{18}O. The aldehyde carbonyl groups of (R) and (S)-lactaldehyde were also labeled with ^{18}O by equilibration in H$_2$^{18}O. Reduction of each of the ^{18}O-enriched enantiomers with LiAlH$_4$ then yielded samples of (R)-1,2-propanediol-1-^{18}O and (S)-1,2-propanediol-1-^{18}O. The syntheses of these three ^{18}O-enriched propanediol substrates are summarized in Figure VI-60.

Rétey et al. then treated each of the ^{18}O propanediol substrates just described with diol dehydrase and used mass spectrometry to determine the amount of ^{18}O isotope retained in the propionaldehyde product. Naturally any ^{18}O initially present in the propionaldehyde products must reside in the aldehyde carbonyl, and the syntheses outlined in Figure VI-60 emphasize that the carbonyl oxygen of an aldehyde will undergo exchange reactions with the oxygen atoms of water. To minimize the loss of ^{18}O present in the initially formed propionaldehyde product, the diol dehydrase incubations were carried out in the additional presence of NADH and alcohol dehydro-

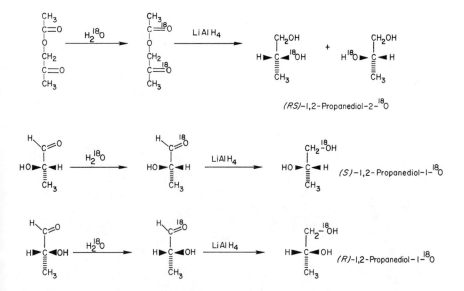

Figure VI-60. Synthesis of ^{18}O-enriched propane-1,2-diol substrates.

genase. The second-mentioned enzymatic reaction rapidly converted the propionaldehyde product of the diol dehydrase reaction into 1-propanol. In this way the exchangeable carbonyl oxygen of the aldehyde product was rapidly trapped as a nonexchangeable hydroxyl oxygen before extensive loss of the ^{18}O label into the water could occur. The experimental results obtained in these experiments are summarized and interpreted mechanistically in Figure VI-61.

It was found that the propanol product isolated after incubation of the racemic 1,2-propanediol-2-^{18}O substrate retained 43% of the original ^{18}O. This clearly indicates a molecular rearrangement of the hydroxyl oxygen at C-2 to the C-1 position of the initially formed propionaldehyde product. This rearrangement process may be interpreted in terms of a 1,2-epoxide intermediate as outlined in Figure VI-61. The propanol product isolated from the (S)-1,2-propanediol-1-^{18}O incubation retained 88% of the original isotopic label, whereas the propanol from the (R)-1,2-propanediol-1-^{18}O incubation retained only 8% of the original isotopic label. These results eliminate the possibility that retention or loss of the oxygen label in the propanediol during the rearrangement is determined by the position of the original label (oxygen attached to C-1 or C-2). The retention and loss of the ^{18}O label displays a distinct stereochemical dependence. The results are most easily rationalized in terms of an initial propane-1,1-diol product,

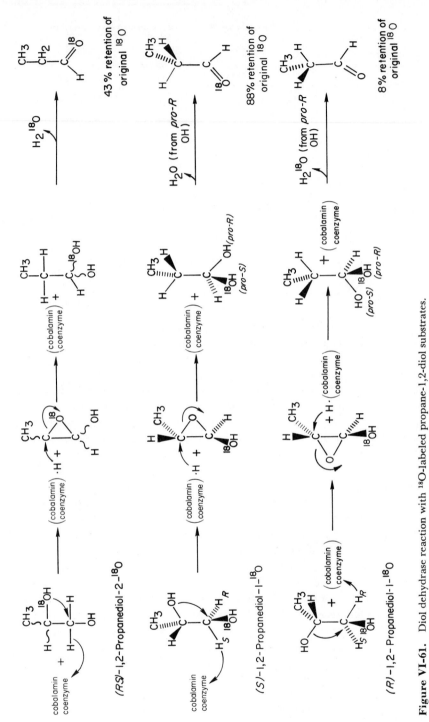

Figure VI-61. Diol dehydrase reaction with ^{18}O-labeled propane-1,2-diol substrates.

which then undergoes an enzymatically catalyzed dehydration to yield the propionaldehyde. Although the hydroxyl functions of propane-1,1-diol molecule are chemically like and would be randomly lost in a chemical dehydration, the reader will immediately recognize that these hydroxyls comprise geometrically distinct, enantiotopic, paired functions. Therefore he would expect an enzymatic dehydration process to differentiate between these paired C-1 hydroxyls on the basis of the differing absolute geometries. Since the hydroxyl groups at C-2 of the propane-1,2-diols that rearrange onto the C-1 position *have opposite configurations in the two enantiomeric substrates*, an enzymatic dehydration of the propane-1,1-diols which maintains a *constant absolute stereospecificity will result in the loss of the original C-1 hydroxyl oxygen in the reaction of one substrate and in the retention of the original C-1 hydroxyl oxygen in the reaction of the enantiomeric substrate* (see Figure VI-61). The reaction sequence proposed in Figure VI-61, therefore, can explain both the C-2 to C-1 oxygen migration and the stereochemically dependent loss of the C-1 oxygen observed by Rétey and co-workers.

We have yet another clear illustration of the principle that biological differentiation between chemically like, paired groups is based on differing absolute geometries because the fate of the original C-1 oxygen from the propane-1,2-diol substrates is stereochemically dependent. In terms of the topic of this book, it is also interesting to note that the biological stereospecificity detected by Rétey and co-workers can now be cited as evidence for an enzymatically catalyzed dehydration of a propane-1,1-diol intermediate. This present acceptance, indeed anticipation, of biological differentiation between chemically like, enantiotopic paired groups is a measure of the growth of our biochemical knowledge. Obviously the present situation contrasts markedly with the scientific atmosphere in 1941, when Evans and Slotin and Wood et al. first detected biological stereospecificity toward enantiotopic paired groups in the operations of the TCA cycle (cf. Chapter I).

The chemical reduction catalyzed by the cobalamin-containing ribonucleotide reductase from *Lactobacillus leichmannii* also displays a stereochemically interesting feature. This enzyme catalyzes the conversion of a ribonucleotide triphosphate to the 2'-deoxyribonucleotide triphosphate. As illustrated in Figure VI-62, it has been established that the reduction occurs with retention of configuration at the C-2' carbon atom of the ribose moiety.[57,58] The original (*R*) C-2' hydroxyl group is replaced by a hydrogen that is derived from NADPH (but that undergoes exchange with protons in the solvent during the reaction) and becomes bonded to the C-2' carbon in the *pro-R* orientation. The mechanism responsible for this stereochemical result has not been established.

Although the biological stereospecificities associated with the reactions catalyzed by the various cobalamin-requiring enzymes are extremely inter-

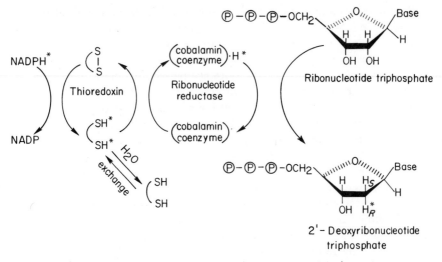

Figure VI-62.   Ribonucleotide reductase (cobalamin-requiring) reaction.

esting, our major reason for choosing the cobalamin coenzyme systems as the final instructive example of this book involves still another stereochemical feature of these reactions. The coenzyme present in each of the enzymatic reactions described previously is the 5′-deoxyadenosyl cobalamin structure pictured in Figure VI-63. Furthermore, although the diol dehydrase process as outlined in Figure VI-61 implies an intramolecular shift of a hydrogen from C-1 to C-2, in fact, experiments performed in Abeles' laboratory established that only about 1% of a 1-tritiopropane-1,2-diol substrate undergoes a true intramolecular rearrangement to yield a 2-tritopropionaldehyde molecule.[59] The overwhelming majority of the 1-tritiopropane-1,2-diol molecules react by transferring the C-1 tritium label to an *additional* competing substrate molecule. It is not known whether other cobalamin-enzyme-catalyzed rearrangements involve such intermolecular transfers. It has been determined, however, that during the methylmalonyl CoA mutase,[60] the methylaspartate mutase,[61] and the diol dehydrase reaction,[59] *hydrogens are transferred from the substrate to the 5′-deoxyadenosyl cobalamin coenzyme and from the cobalamin coenzyme into the reaction products.* In the case of the ribonucleotide reductase, the hydride equivalents derived from NADPH are transferred via sulfhydryl groups on the thioredoxin factor and exchange with solvent at this stage (cf. Figure VI-62). Therefore, a transfer of hydrogen from the cobalamin coenzyme to the product is not experimentally observed in the ribonucleotide reductase reaction. Instead, an enzyme-dependent exchange of hydrogens of the coenzyme with hydrogens of the aqueous solvent is

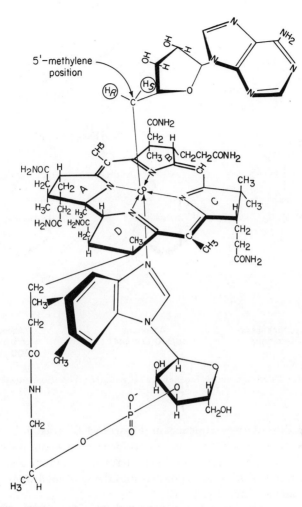

Figure VI-63.   5′-Deoxyadenosyl cobalamin coenzyme.

detected.[62,63] In all cases, however, the hydrogen atoms involved in the transfer and exchange processes are attached to the 5′-carbon of the 5′-deoxyadenosyl moiety of the coenzyme. These hydrogens are specially indicated in Figure VI-63.

This exchange process has been most extensively investigated in the diol dehydrase system. When 1-tritiopropane-1,2-diol was incubated with 5′-deoxyadenosyl cobalamin in the presence of diol dehydrase, *two atoms* of tritium were incorporated into *each molecule of the cobalamin coenzyme* struc-

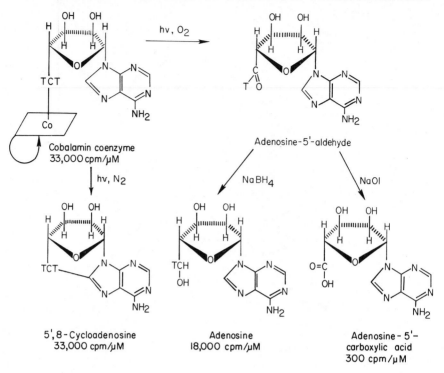

Figure VI-64. Degradation of biochemically tritiated 5'-deoxyadenosyl cobalamin coenzyme.

ture.[64] The degradative sequence in Figure VI-64 was then utilized to establish the position of the tritium atoms incorporated into the cobalamin. As Figure VI-64 shows, the experimental results demonstrated that both atoms of tritium had been incorporated into the C-5' methylene position of the 5'-deoxyadenosyl moiety.

The diol dehydrase-catalyzed transfer of label from the C-5' position of the cobalamin coenzyme to the propionaldehyde product was investigated with tritium-labeled cobalamins prepared by chemical synthesis.[59] Adenosine, protected as a 2',3'-isopropylidene derivative, was oxidized to the 5'-formyl derivative and then reduced back to 2',3'-isopropylideneadenosine with sodium borotritide. As outlined in Figure VI-65, this procedure produced an adenosine derivative labeled with tritium on the 5'-methylene position. The tritium-labeled isopropylideneadenosine was then converted into the 5'-tosylate ester and condensed with reduced vitamin B_{12} (B_{12s}). This reaction produces a Co–C-5' bond and thus yields 5'-tritio-5'-deoxy-2',3'-isopropylideneadenosyl cobalamin, as shown in Figure VI-65. Mild acid

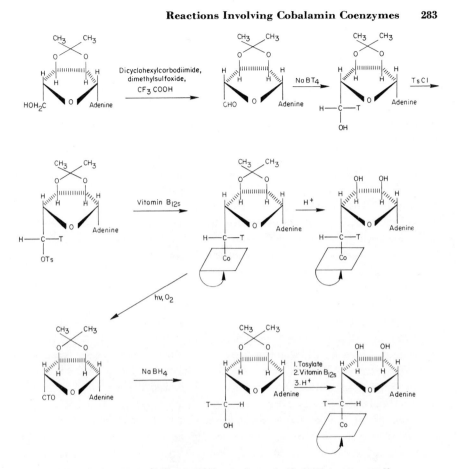

Figure VI-65. Synthesis of 5′-tritio-5′-deoxyadenosyl cobalamin coenzyme.[59]

hydrolysis of the protecting ketal grouping then freed the 5′-tritium-labeled cobalamin coenzyme necessary for the investigation. When this tritium-labeled cobalamin coenzyme was incubated with unlabeled propane-1,2-diol and the diol dehydrase enzyme, the *entire* tritium label originally present in the coenzyme was transferred to the propionaldehyde product.

Although the 5′-formyladenosine derivative reduced with sodium borotritide in the reaction sequence is a chiral molecule, the foregoing chemical reduction would be expected to display only a slight degree of stereoselectivity. Therefore, both of the 5′-methylene hydrogens of the 2′,3′-isopropylideneadenosine product should have been partially labeled with tritium by this chemical synthesis. Because of the vital importance of this stereochemical feature, however, the chemical synthesis was also carried out by reducing the 5'-tritioformyladenosine derivative with unlabeled sodium borohydride.[64] If the sodium borohydride reduction had displayed an unexpected degree of chemical stereoselectivity, then the 5'-tritium label in this second adenosine derivative would necessarily have possessed the opposite absolute stereochemistry of the 5'-tritium label in the first 5'-tritioadenosine sample. As we see in Figure VI-65, the second 5'-tritioadenosine derivative was also converted to the cobalamin coenzyme form. When the second synthetic 5'-tritio-5'-deoxyadenosyl cobalamin coenzyme sample was incubated with propane-1,2-diol in the presence of diol dehydrase, *again* it was observed that the *entire* tritium label was transferred from the labeled coenzyme to the propionaldehyde product.

These experimental results established that, during the diol dehydrase reaction, tritium is transferred from the 1-tritiopropane-1,2-diol substrate *to both* the 5'-methylene hydrogens of the 5'-deoxyadenosyl cobalamin and then *from both* the 5'-methylene hydrogens of the cobalamin coenzyme to C-2 of the propionaldehyde product. An inspection of the 5'-deoxyadenosyl cobalamin structure of Figure VI-63, however, immediately reveals that the methylene hydrogens at the C-5' position are nonequivalent. The C-5' position, as indicated, is a *pro*-chiral center, bearing diastereotopic paired hydrogens. Once again, nmr spectroscopic studies emphasize the chemical and geometrical nonequivalence of these diastereotopic paired hydrogen atoms. Separate resonances for the two C-5' paired hydrogens of 5'-deoxyadenosyl cobinamide have been recently observed and identified.[65]

The major theme of this book, developed both by citing experimental examples and by analyzing the stereochemical principles of enzymatic catalysis, is that biological stereospecificity between nonequivalent paired groups will be manifest in biochemical processes. The results of the foregoing investigations seem inconsistent with this carefully developed theme. Because the experiments just described established only the final distribution of label, however, a potential explanation is that the reaction involves a series of steps in which the overall labeling result appears to lack biological stereospecificity, although, in fact, the nonequivalent paired hydrogens at C-5' of the cobalamin coenzyme remain *kinetically distinct*. For example, the exchange and transfer process associated with the diol dehydrase reaction could proceed via a sequential "merry-go-round" as suggested in Figure VI-66. In such a mechanism each step is stereospecific and only one of the nonequivalent methylene hydrogens at C-5' would be transferred to the initially formed propionaldehyde product. If an inversion at the C-5' position were to occur in the initial reaction, however, the subsequent reaction of this cobalamin coenzyme to yield a second molecule of propionaldehyde would

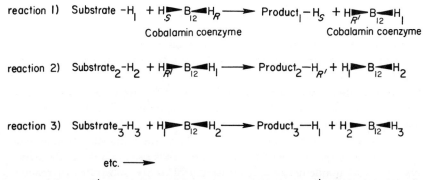

(H$_{R'}$= H$_S$ 5'-cobalamin hydrogen that was initially H$_R$ 5'-cobalamin hydrogen)

Figure VI-66. Possible stereospecific "merry-go-round" mechanism of cobalamin coenzyme-catalyzed rearrangements.

involve the stereospecific transfer of that paired methylene hydrogen which originally possessed the opposite absolute configuration. Finally, as depicted in Figure VI-66, the third reaction of this cobalamin coenzyme molecule would involve the stereospecific transfer of a hydrogen from the C-5' position that was derived from the initial reaction with a propane-1,2-diol substrate molecule. It is possible to formulate several such "merry-go-round" processes, lacking overall stereospecificity, even though each individual transfer is stereospecific and the transferring hydrogens remain kinetically distinct.

Because the hydrogen stereospecifically transferred to the initially formed product molecule can be kinetically differentiated from that transferred to subsequently formed product molecules, the "merry-go-round" possibilities are subject to experimental test. Miller and Richards carried out an appropriate series of experiments with trideuteriomethylmethylmalonyl CoA substrates and the methylmalonyl CoA mutase enzyme.[66] The distribution of deuterium in the succinyl CoA products, when analyzed by computer methods, indicated that the hydrogen abstracted from the methyl group of the methylmalonyl CoA substrate becomes *one of three equivalent coenzyme hydrogens* before a hydrogen is transferred to regenerate methylmalonyl CoA or to yield succinyl CoA.

Essenberg et al. have also investigated the kinetics of the hydrogen transfer processes catalyzed by the diol dehydrase enzyme.[67] In an extensive set of experiments, the rate of tritium transfer from C-1 of (RS)-propane-1,2-diol to the C-5' position of the cobalamin coenzyme, the rate of tritium transfer from C-5' of the coenzyme to the C-2 position in the propionaldehyde product, and the overall rate of tritium transfer from the propanediol substrate to the propionaldehyde product were measured. The measured rates of these transfer reactions were found to be consistent with a reaction sequence in which the B_{12} coenzyme functions as an obligatory intermediate hydrogen carrier between the propanediol substrate and the propionaldehyde product.

In addition, the rates of tritium transfer from 5′-tritiocobalamin coenzyme to products were compared with deuterium-labeled substrates and with unlabeled substrates. The transfer of tritium from the C-5′ position of the cobalamin coenzyme to the product was enhanced 18-fold when 1,1,2,2-tetradeuterioethylene glycol was the substrate, relative to the tritium transfer with nondeuterated ethylene glycol as the substrate. In reactions involving both deuterium-labeled and nondeuterium-labeled substrates [e.g., (RS)-1,1-dideuteriopropane-1,2-diol and ethylene glycol] it was found that the presence of deuterium in a substrate molecule enhanced the transfer of tritium from the coenzyme to the product derived from the deuterated substrate by four to sevenfold. An evaluation of the kinetic isotope effects $(k_H:k_D:k_T)$ suggested that the enhanced probabilities of tritium transfers from the C-5′ position of the cobalamin coenzyme in reactions involving deuterium-labeled substrates resulted from an *intermediate involving three equivalent hydrogens*—one contributed by the substrate and two derived from the C-5′ methylene hydrogens—and that *in any one turnover, any of the three equivalent hydrogens could be transferred to the reaction product*.

Thus the experimental evidence available indicates that the substrate hydrogen undergoing rearrangement and the two nonequivalent diastereotopic C-5′ methylene hydrogens of the cobalamin coenzyme become stereochemically *and* kinetically indistinguishable during the course of the mutase and dehydrase reactions. Based on such findings, the sequence of steps outlined in Figure VI-67 has been proposed for the rearrangement reactions catalyzed by the cobalamin-requiring enzymes.[66,67] The postulated 5′-deoxyadenosine intermediate* can satisfactorily explain the observed lack of biological stereospecificity toward the nonequivalent, paired C-5′ methylene hydrogens in these reactions. These hydrogens become part of a methyl group and, by virtue of rotational superpositioning, the methyl hydrogens are equivalent. In addition, because one of the methyl hydrogens of the 5′-deoxyadenosine is derived from the rearranging hydrogen of the substrate, this proposed mechanism will explain how the rearranging substrate hydrogen and the two original C-5′ methylene hydrogens of the coenzyme become three kinetically indistinguishable hydrogens during the enzymatic reactions.

---

* A 5′-deoxyadenosine intermediate has also been postulated in the ribonucleotide reductase reaction (cf. Ref. 63) and in the ethanolamine ammonia-lyase reaction (cf. Ref. 68).

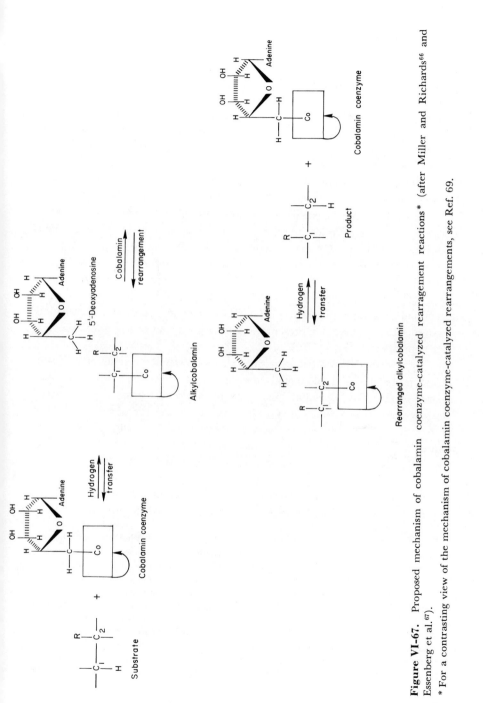

Figure VI-67. Proposed mechanism of cobalamin coenzyme-catalyzed rearrangement reactions* (after Miller and Richards[66] and Essenberg et al.[67]).

* For a contrasting view of the mechanism of cobalamin coenzyme-catalyzed rearrangements, see Ref. 69.

It must be emphasized that the sequence of steps in Figure VI-67 was proposed despite observations that formation of the cobalt–carbon bond in the cobalamin coenzyme structure required stringently controlled chemical or biochemical conditions (cf. Figure VI-65), that the known reactions which resulted in the cleavage of the cobalt–carbon bond of the cobalamin coenzymes led to loss of the coenzyme activity, and that 5′-deoxyadenosine had not been identified as an intermediate in the mutase or dehydrase reactions.* In short, the unique mechanism in Figure VI-67 was proposed on the basis of the observed lack of biological stereospecificity at the C-5′ methylene position, despite the absence both of traditional chemical evidence for such a process and of a known chemical analogy for such a series of steps.

In 1941 a failure to recognize the possibility of biological stereospecificity between paired groups led investigators to "eliminate" citric acid as a potential intermediate in TCA cycle metabolism. By 1969 the principle of biological stereospecificity between nonequivalent paired groups was so widely accepted that a unique biochemical mechanism could be proposed on the basis of stereochemical observations. For this reason, the rearrangement reactions catalyzed by the cobalamin-requiring enzymes are appropriate examples to conclude our presentation of substrate symmetry and biological stereospecificity.

By stressing the relationship that exists between the symmetry of substrate molecules and biological stereospecificity between chemically like, paired groups, we hope that this aspect of biochemistry has been rendered less mysterious—less magical. We may still marvel at the phenomenon, but it can nevertheless be understood and predicted in terms of the three-dimensional substrate-enzyme interactions characteristic of biological catalysis. We also hope that the many examples of biological stereospecificity between chemically like, paired groups described in this book will dispel the illusion that this phenomenon is uncommon. It should be stressed that the examples cited here are far from encyclopedic and are only meant to be

---

* 5′-Deoxyadenosine had been tentatively identified as a nucleoside product when the propanediol dehydrase-cobalamin coenzyme complex was inactivated by noncatalytic reaction with either oxygen or glycolaldehyde.[70] Only recently, however, has evidence of the formation of 5′-deoxyadenosine during the catalytic functioning of a cobalamin coenzyme been obtained. Babior found that 5′-deoxyadenosine is formed when 5′-deoxyadenosyl cobalamin coenzyme and the ethanolamine ammonia-lyase enzyme are incubated in the presence of substrate (ethanolamine).[68] It was also found that, in the absence of substrate, the enzyme catalyzed a cleavage of the cobalt–carbon bond of the coenzyme to produce a mixture of nucleoside products, including adenosine-5′-aldehyde. The mechanism for the cobalamin coenzyme-requiring ethanolamine ammonia-lyase reaction proposed by Babior to account for these observations is similar to that outlined in Figure VI-67.

representative of the numerous manifestations of biological stereospecificity between paired groups that have been investigated and described.

Finally, we hope the discussions of this book have indicated how investigations of biological differentiation between paired groups can provide insights into mechanistic details of biochemical reactions. It seems clear that as our knowledge of biochemical catalysis becomes more detailed, an understanding of the phenomenon of biological stereospecificity between chemically like, paired groups will become ever more important. We hope that this book will contribute to such an understanding.

REFERENCES

1. K. R. Hanson and I. A. Rose, *Proc. Natl. Acad. Sci. (U.S.)*, **50**, 981 (1963).

2. S. Englard, *J. Biol. Chem.*, **235**, 1510 (1960).

3. G. E. Lienhard and I. A. Rose, *Biochemistry*, **3**, 185 (1964).

4. T. T. Tchen and H. van Milligan, *J. Amer. Chem. Soc.*, **82**, 4115 (1960).

5. A. J. Kluyver and A. G. J. Boezaardt, *Rec. Trav. Chim. Pays-Bas*, **58**, 956 (1939).

6. T. Posternak, *Helv. Chim. Acta*, **29**, 1991 (1946).

7. B. Magasanik and E. Chargaff, *J. Biol. Chem.*, **174**, 173 (1948).

8. M. Antia, D. S. Hoare, and E. Work, *Biochem. J.*, **65**, 448 (1957).

9. J. W. Cornforth, J. W. Redmond, H. Eggerer, W. Buckel, and C. Gutschow, *Nature*, **221**, 1212 (1969).

10. J. Lüthy, J. Rétey, and D. Arigoni, *Nature*, **221**, 1213 (1969).

11. H. Eggerer, W. Buckel, H. Lenz, P. Wunderwald, G. Gottschalk, J. W. Cornforth, C. Donninger, R. Mallaby, and J. W. Redmond, *Nature*, **226**, 517 (1970).

12. J. Rétey, J. Lüthy, and D. Arigoni, *Nature*, **226**, 519 (1970).

13. J. Rétey, J. Seibl, D. Arigoni, J. W. Cornforth, G. Ryback, W. P. Zeylemaker, and C. Veeger, *Eur. J. Biochem.*, **14**, 232 (1970).

14. C. G. Swain, E. C. Stivers, J. F. Reuwer, Jr., and L. J. Schaad, *J. Amer. Chem. Soc.*, **80**, 5885 (1958).

15. J. W. Cornforth, J. W. Redmond, H. Eggerer, W. Buckel, and C. Gutschow, *Eur. J. Biochem.*, **14**, 1 (1970).

16. J. Bové, R. O. Martin, L. L. Ingraham, and P. K. Stumpf, *J. Biol. Chem.*, **234**, 999 (1959).

17. H. Eggerer, *Biochem. Z.*, **343**, 111 (1965).

18. G. Popják and J. W. Cornforth, *Biochem. J.*, **101**, 553 (1966).

19. G. Popják, J. W. Cornforth, R. H. Cornforth, R. Ryhage, and D. S. Goodman, *J. Biol. Chem.*, **237**, 56 (1962).

20. D. Arigoni, *Experientia*, **14**, 153 (1958).

21. A. J. Birch, M. Kocor, N. Sheppard, and J. Winter, *J. Chem. Soc.*, 1502 (1962).

22. R. U. Lemieux and J. Howard, *Can. J. Chem.*, **41**, 308 (1963).

23. C. Donninger and G. Ryback, *Biochem. J.*, **91**, 11p (1964).

24. C. Donninger and G. Popják, *Proc. Royal Soc. Ser. B*, **163**, 465 (1966).

25. I. A. Rose, *J. Biol. Chem.*, **245**, 6052 (1970).

26. H. F. Fisher, E. E. Conn, B. Vennesland, and F. H. Westheimer, *J. Biol. Chem.*, **202**, 687 (1953); F. A. Loewus, F. H. Westheimer, and B. Vennesland, *J. Amer. Chem. Soc.*, **75**, 5018 (1953); H. R. Levy, F. A. Loewus, and B. Vennesland, *ibid.*, **79**, 2949 (1957).

27. F. H. Westheimer, H. F. Fisher, E. E. Conn, and B. Venneslana, *J. Amer. Chem. Soc.*, **73**, 2403 (1951).

28. M. E. Pullman, A. San Pietro, and S. P. Colowick, *J. Biol. Chem.*, **206**, 129 (1954).

29. R. Bentley, *Molecular Asymmetry in Biology*, Vol. II, Academic Press, 1970, Table I, pp. 6–9; Table II, p. 11.

30. J. W. Cornforth, R. H. Cornforth, C. Donninger, G. Popják, G. Ryback, and G. J. Schroepfer, Jr., *Proc. Royal Soc. Ser. B*, **163**, 436 (1966).

31. M. W. McDonough and W. A. Wood, *J. Biol. Chem.*, **236**, 1220 (1961).

32. I. A. Rose and E. L. O'Connell, *Biochim. Biophys. Acta*, **42**, 159 (1960).

33. C. O. Clagett, N. E. Tolbert, and R. H. Burris, *J. Biol. Chem.*, **178**, 977 (1949); I. Zelitch and S. Ochoa, *J. Biol. Chem.*, **201**, 707 (1953).

34. I. A. Rose, *J. Amer. Chem. Soc.*, **80**, 5835 (1958).

35. C. K. Johnson, E. J. Cabe, M. R. Taylor, and I. A. Rose, *J. Amer. Chem. Soc.*, **87**, 1802 (1965).

36. Y. J. Topper, *J. Biol. Chem.*, **225**, 419 (1957).

37. M. Cohn, J. E. Pearson, E. L. O'Connell, and I. A. Rose, *J. Amer. Chem. Soc.*, **92**, 4095 (1970).

38. I. A. Rose and E. L. O'Connell, *J. Biol. Chem.*, **236**, 3086 (1961).

39. H. Simon and R. Medina, *Z. Naturforsch.*, **21b**, 496 (1966).

40. I. A. Rose, E. L. O'Connell, and R. P. Mortlock, *Biochim. Biophys. Acta*, **178**, 376 (1969).

41. S. V. Rieder and I. A. Rose, *J. Biol. Chem.*, **234**, 1007 (1959).

42. J. E. D. Dyson and E. A. Noltmann, *J. Biol. Chem.*, **243**, 1401 (1968).

43. B. Bloom and Y. J. Topper, *Science*, **124**, 982 (1956).

44. B. L. Horecker, P. T. Rowley, E. Grazi, T. Cheng, and O. Tchola, *Biochem. Z.* **338**, 36 (1963).

45. R. D. Kobes, R. T. Simpson, B. L. Vallee, and W. J. Rutter, *Biochemistry*, **8**, 585 (1969).

46. M. Akhtar and P. M. Jordan, *Tetrahedron Lett.*, 875 (1969).

47. M. Akhtar and P. M. Jordan, *Chem. Commun.*, 1691 (1968).

48. M. Sprecher, M. J. Clark, and D. B. Sprinson, *J. Biol. Chem.*, **241**, 872 (1966).

49. P. Overath, G. M. Kellerman, F. Lynen, H. P. Fritz, and H. J. Keller, *Biochem. Z*, **335**, 500 (1962).

50. J. D. Erfle, J. M. Clark, Jr., R. F. Nystrom, and B. C. Johnson, *J. Biol. Chem.*, **239**, 1920 (1964).

51. M. Sprecher, R. L. Switzer, and D. B. Sprinson, *J. Biol. Chem.*, **241**, 864 (1966).

52. A. A. Iodice and H. A. Barker, *J. Biol. Chem.*, **238**, 2094 (1963).

53. T. Yamane, T. Kato, S. Shimizu, and S. Fukui, *Arch. Biochem. Biophys.*, **113**, 362 (1966).

54. B. Zagalak, P. A. Frey, G. L. Karabatsos, and R. H. Abeles, *J. Biol. Chem.*, **241**, 3028 (1966).

55. J. Rétey, A. Umani-Ronchi, and D. Arigoni, *Experientia*, **22**, 72 (1966).

56. J. Rétey, A. Umani-Ronchi, J. Seibl, and D. Arigoni, *Experientia*, **22**, 503 (1966).

57. T. J. Batterham, R. K. Ghambeer, R. L. Blakley, and C. Brownson, *Biochemistry*, **6**, 1203 (1967).

58. C. E. Griffin, F. D. Hamilton, S. P. Hopper, and R. Abrams, *Arch. Biochem. Biophys.*, **126**, 905 (1968).

59. P. A. Frey, M. K. Essenberg, and R. H. Abeles, *J. Biol. Chem.*, **242**, 5369 (1967).

60. J. Rétey and D. Arigoni, *Experientia*, **22**, 783 (1966).

61. R. L. Switzer, B. G. Baltimore, and H. A. Barker, *J. Biol. Chem.*, **244**, 5263 (1969).

62. W. S. Beck, R. H. Abeles, and W. G. Robinson, *Biochem. Biophys. Res. Commun.*, **25**, 421 (1966).

63. H. P. C. Hogenkamp, R. K. Ghambeer, C. Brownson, R. L. Blakley, and E. Vitols, *J. Biol. Chem.*, **243**, 799 (1968).

64. P. A. Frey, S. S. Kerwar, and R. H. Abeles, *Biochem. Biophys. Res. Commun.*, **29**, 873 (1967).

65. S. A. Cockle, H. A. O. Hill, R. J. P. Williams, B. E. Mann, and J. M. Pratt, *Biochim. Biophys. Acta*, **215**, 415 (1970).

66. W. W. Miller and J. H. Richards, *J. Amer. Chem. Soc.*, **91**, 1498 (1969).

67. M. K. Essenberg, P. A. Frey, and R. H. Abeles, *J. Amer. Chem. Soc.*, **93**, 1242 (1971).

68. B. M. Babior, *J. Biol. Chem.*, **245**, 6125 (1970).

69. G. N. Schrauzer, R. J. Holland, and J. A. Seck, *J. Amer. Chem. Soc.*, **93**, 1503 (1971).

70. O. W. Wagner, H. A. Lee, Jr., P. A. Frey, and R. H. Abeles, *J. Biol. Chem.*, **241**, 1751 (1966).

71. K. Clifford, J. W. Cornforth, R. Mallaby, and G. T. Phillips, *Chem. Commun.*, 1599 (1971).

BIBLIOGRAPHY

BIOLOGICAL STEREOSPECIFICITY

"A Nonenzymatic Illustration of 'Citric Acid Type' Asymmetry: The *Meso*-Carbon Atom," P. Schwartz and H. E. Carter, *Proc. Natl. Acad Sci.* (*U.S.*), **40**, 499 (1954).

"The Nature of Substrate Asymmetry in Stereoselective Reactions," H. Hirschmann, *J. Biol. Chem.*, **235**, 2762 (1960).

"The Steric Course of Enzymatic Reactions at *Meso* Carbon Atoms: Application of Hydrogen Isotopes," H. R. Levy, P. Talalay, and B. Vennesland, in *Progress in Stereochemistry*, Vol. 3, P. B. D. de la Mare and W. Klyne, Eds., Butterworths, 1962, Chapter 8.

"Newer Aspects of Enzymatic Stereospecificity," H. Hirschmann, in *Comprehensive Biochemistry*, Vol. 12, M. Florkin and E. H. Stotz, Eds., Elsevier, 1964, Chapter 7.

"Applications of the Sequence Rule. I. Naming the Paired Ligands g,g at a Tetrahedral Atom X_{ggij}. II. Naming the Two Faces of a Trigonal Atom Y_{ghi}," K. R. Hanson, *J. Amer. Chem. Soc.*, **88**, 2731 (1966).

"Chirality Due to the Presence of Hydrogen Isotopes at Noncyclic Positions," D. Arigoni and E. L. Eliel, in *Topics in Stereochemistry*, Vol. 4, E. L. Eliel and N. L. Allinger, Eds., Wiley, 1969, p. 127

"The Chiral Methyl Group—Its Biochemical Significance," J. W. Cornforth, *Chemistry in Britain*, **6**, 431 (1970).

"Stereochemistry of Enzymatic Reactions at Prochiral Centers," H. G. Floss, *Naturwiss.*, **57**, 435 (1970).

Molecular Asymmetry in Biology, Vols. I and II, R. Bentley, Academic Press, 1969, 1970. Bentley's two-volume set is a detailed and quite comprehensive treatment of most of the topics presented in this book. These two volumes could be most profitably consulted when the concepts and information introduced in this book have been mastered.

The selected references cited here treat some of the topics of this book in greater depth. They are suggested as initial additional references for those individuals who wish to proceed beyond the level of this presentation.

MOLECULAR SYMMETRY AND STEREOCHEMISTRY

Introduction to Stereochemistry, K. Mislow, W. A. Benjamin, 1966.
Stereochemistry of Carbon Compounds, E. L. Eliel, McGraw-Hill, 1962.
Symmetry in Chemistry, H. H. Jaffé and M. Orchin, Wiley, 1965.
Chemical Applications of Group Theory, F. A. Cotton, Wiley, 1963.
"Symmetry, Point Groups, and Character Tables," M. Orchin and H. H. Jaffé, a
    series of resource papers in *J. Chem. Educ.*, **47**, 246, 372, 510 (1970).

CHEMICALLY LIKE, PAIRED GROUPS AND EQUIVALENT NUCLEI

"Stereochemical Relationships of Groups in Molecules," K Mislow and M. Raban,
    in *Topics in Stereochemistry*, Vol. I, N. L. Allinger and E. L. Eliel, Eds., Wiley,
    1967, p. 1.
"Equivalence of Nuclei in High-Resolution Nuclear Magnetic Resonance Spectros-
    copy," M. van Gorkom and G. E. Hall, *Quart. Rev. (London)*, **22**, 14 (1968).

R AND S CONFIGURATIONAL NOMENCLATURE

"An Introduction to the Sequence Rule," R. S. Cahn, *J. Chem Educ.*, **41**, 116 (1964).
"The Specification of Asymmetric Configuration in Organic Chemistry," R. S. Cahn,
    C. K. Ingold, and V. Prelog, *Experientia*, **12**, 81 (1956).
"Specification of Molecular Chirality," R. S. Cahn, C. K. Ingold, and V. Prelog,
    Angew. Chem. Int. Ed. Engl., **5**, 385 (1966).

PROTEIN STRUCTURE AND ACTION

"Structure, Function and Evolution in Proteins," Brookhaven Symposia in Biology,
    No. 21 (1968), pt. 1 and 2, BNL-50116.
The Structure and Action of Proteins, R. E. Dickerson and I. Geis, Harper and Row, 1969.
"Implications of the X-Ray Crystallographic Studies of Protein Structure," L.
    Stryer, *Annu. Rev. Biochem.*, **37**, 25 (1968).
"X-Ray Diffraction Studies of Enzymes," D. M. Blow and T. A. Steitz, *Annu. Rev.
    Biochem.*, **39**, 101 (1970).

AUTHOR INDEX

Carrell, H. L., 107 (IV-13)
Carter, H. E., 13 (I-16); 78 (III-15); 112–113, 121, 122, 134 (V-5)
Caserio, M. C., 35n
Cava, M. P., 98n
Chargaff, E., 186–188 (VI-7)
Cheng, T., 262 (VI-44)
Chowdhury, A. A., 158 (V-29)
Ciganek, E., 100 (IV-4)
Clagett, C. O., 246n (VI-33)
Clark, J. M. Jr., 272n (VI-50)
Clark, M. J., 271–272 (VI-48)
Clifford, K., 234, 271n (VI-71)
Cockle, S. A., 284 (VI-65)
Cohen, J. A., 164 (V-46)
Cohen, S. G., 89; 164–174 (V-53, V-57)
Cohn, M., 250–253 (VI-37)
Colowick, S. P., 131 (V-19); 241 (VI-28)
Conn, E. E., 15–16 (II-2); 240 (VI-27); 240–241, 243 (VI-26)
Cooper, T. G., 147–149 (V-23)
Cope, A. C., 100 (IV-4)
Cornforth, J. W., 194, 196–198, 204–211 (VI-9, VI-11); 204 (VI-13); 206n (VI-15); 212–233 (VI-18, VI-19); 234, 271n (VI-71); 243 (VI-30)
Cornforth, R. H., 212, 214–215 (VI-19); 243 (VI-30)

Daeniker, H. U., 98n
DeSa, R. J., 164 (V-44)
Dickens, F., 141 (V-22)
Dickman, S. R., 104 (IV-9)
Dittbrenner, S., 158 (V-29)
Djerassi, C., 142n
Donninger, C., 194, 196–198, 204–211 (VI-11); 217 (VI-23); 219n (VI-24); 243 (VI-30)
Dummel, R. J., 104–109 (IV-12, IV-13), 126 (V-18)
Dyson, J. E. D., 255n (VI-42)

Eggerer, H., 194, 196–198, 204–211 (VI-9, VI-11); 206n (VI-15); 211 (VI-17)
Elwyn, D., 115–116, 162 (V-6)
Englard, S., 123–126, 154 (V-11); 131 (V-19); 183 (VI-2)
Erfle, J. D., 272n (VI-50)
Essenberg, M. K., 280, 282–284 (VI-59); 285–287 (VI-67)

Evans, E. A. Jr., 10 (I-8); 123 (V-9)

Faller, L., 164(V-40)
Fanshier, D. W., 104–109 (IV-10, IV-11); 126 (V-18)
Farrar, T. C., 124n (V-14)
Fischer, E., 50–55, 57
Fisher, H. F., 240 (VI-27); 240–241, 243 (VI-26)
Fondy, T. P., 123–124, 150 (V-12)
Frey, P. A., 273–176 (VI-54); 280, 282–284 (VI-59); 282–284 (VI-64); 285–287 (VI-67); 288n (VI-70)
Fritz, H. P., 272n (VI-49)
Fukui, S., 273n (VI-53)

Gawron, O., 123–124, 150 (V-12)
Gay-Lussac, L. J., 2n
Ghambeer, R. K., 279 (VI-57); 281, 286n (VI-63)
Glusker, J. P., 107 (IV-13); 155 (V-24)
Goldberg, S. I., 96–101 (IV-3, IV-5)
Goodman, D. S., 212, 214–215 (VI-19)
Gottschalk, G., 155, 157–158 (V-25, V-26, V-27, V-29); 194, 196–198, 204–211 (VI-11)
Gottwald, L. K., 104–109 (IV-10, IV-11); 126 (V-18)
Grazi, E., 262 (VI-44)
Griffin, E. C., 279 (VI-58)
Gutfreund, H., 164 (V-39)
Gutowsky, H. S., 124n (V-14)
Gutschow, C., 194, 205–208 (VI-9); 206n (VI-15)

Hamilton, F. D., 279 (VI-58)
Hanson, K. R., 13 (I-14); 76n (III-14); 104 (IV-8); 112–113, 135, 143 (V-3); 126, 151–155, 157 (V-15); 183 (VI-1)
Hartley, B. S., 164 (V-38)
Hauser, G., 115–116, 162 (V-6)
Heidelberger, C., 12 (I-12)
Hein, G. E., 172 (V-56)
Helm, D. van der, 68 (III-11); 126 (V-17)
Hemingway, A., 8 (I-7); 10 (I-10); 123 (V-10)
Henderson, R., 164, 174–176 (V-50, V-51, V-52)
Hess, G. P., 164 (V-43, V-44)
Hill, H. A. O., 284 (VI-65)
Himoe, A., 164 (V-43, V-44)

SUBJECT INDEX